Microwave Oven Repair

2nd Edition

To my parents, Orpha and Chester,
and to my wife's parents, Ethel and Richard.

Microwave Oven Repair

2nd Edition

Homer L. Davidson

SECOND EDITION
SECOND PRINTING

Magnameter™ Electronic Systems, 624 Cedar St., Rockford, IL 61102

Library of Congress Cataloging-in-Publication Data

Davidson, Homer L.
    Microwave oven repair / by Homer L. Davidson—2nd ed.
        p.   cm.
    Rev. ed. of: Practical microwave oven repair. 1st ed. 1984.
    Includes index.
    ISBN 0-8306-6457-2 (hard)   ISBN 0-8306-3457-6 (pbk.)
    1. Microwave ovens—Maintenance and repair.   I. Davidson, Homer
L. Practical microwave oven repair.   II. Title.
TX657.064D28   1991
683′.83—dc20   90-24054
                                                               CIP

TAB Books offers software for sale. For information and a catalog, please contact TAB Software Department, Blue Ridge Summit, PA 17294-0850.

Acquisitions Editor: Kimberly Tabor
Book Editor: Barbara Minich
Production: Katherine G. Brown
Book Design: Jaclyn J. Boone
Cover photography by Homer L. Davidson.

Contents

Acknowledgments   **xi**

Introduction   **xii**

1 Some Basic Facts   **I**

Different types of ovens   *1*
Microwave oven frequency   *1*
Fundamental function of the magnetron   *4*
Some dos and don'ts for safety   *7*
Oven installation   *9*
Service technician cautions and warnings   *10*
Basic service precautions   *12*
Checking the outlet receptacle   *13*
Pull the plug   *13*
Lightning protection   *13*
Basic oven components   *17*

2 Required Tools and Test Equipment   **35**

Whirlpool's microwave exposure warnings   *35*
Attaching test equipment   *36*
Basic hand tools   *37*
Volt-ohmmeter   *41*
Digital vom   *41*
DMM diode tests   *42*
DMM transistor tests   *43*

VTVM   *43*
The Magnameter   *44*
Cooking test equipment   *50*
Leakage tests   *53*
The Narda 8100 radiation monitor   *55*
The Simpson 380 series microwave leakage tester   *55*
Panasonic's confirm after repair   *56*

3 Microwave Oven Circuits   57

Different types of fuses   *57*
Low-voltage circuits   *57*
High-voltage circuits   *61*
The controller circuits   *63*
Defrost circuits   *64*
Temperature-control circuits   *66*
Convection circuits   *69*
The cooking cycle   *71*
Removing the wrap or cover   *77*

4 Basic Troubleshooting Procedures   79

Owner's responsibility   *80*
Where to start   *80*
Isolation   *82*
Dead condition   *82*
Oven light on but no cooking   *84*
Oven light on, cook relay energized, no cooking   *87*
No transformer voltage   *89*
Transformer voltage but no cooking   *89*
Typical troubleshooting charts   *90*
Typical oven symptoms and
   possibly defective components   *91*
Proper polarization and proper grounding   *106*
Conclusion   *106*

5 Troubleshooting Microwave Oven Switches   109

Interlock action   *112*
Replacing a defective interlock switch   *115*
Interlock switch replacement   *116*
The monitor switch   *120*
The start and stop switch   *125*
The oven relay   *127*
Relay contact layout   *130*
Timer controls   *134*

Oven lamp switch   *136*
The select switch   *138*

6 Servicing Low-Voltage Problems 139

Low-voltage procedures   *139*
Checking fuses   *141*
Keeps blowing fuses   *141*
Varistor test   *146*
Intermittent oven lamp   *146*
Intermittent oven cooking (cutout)   *148*
Replacing thermal protection   *152*
Oven does not go into cook cycle   *152*
Oven keeps operating (defective triac)   *154*
Reversed triac connections   *157*
Lamp triac tests   *158*
Everything lights up but no cooking (meat probe)   *158*
Different temperature probe tests   *159*
Defective browning unit   *161*
Conclusion   *165*

7 Troubleshooting High-Voltage Problems 167

The magnetron   *167*
The voltage-doubler circuit   *169*
The high-power transformer   *173*
Filament transformer tests   *175*
Arcing hv transformers   *178*
The hv capacitor   *179*
9 volt meter capacitor test   *182*
The hv diode   *182*
How to check the hv circuits   *185*
Microwave oven hot spots   *189*

8 Locating and Replacing the Magnetron 191

How to test   *191*
Checking the magnetron without an hv meter   *194*
Checking the magnetron with the ohmmeter   *196*
Checking the magnetron with a cooking test   *198*
Magnetron failure   *200*
Intermittent thermal unit   *204*
Removing the magnetron   *204*
Keep metal tools away   *207*
Replacing the tube   *207*
Costs too much to fix   *209*
Too hot to handle   *209*
Two explosive examples   *210*

How to order a new tube *210*
How to replace while still under warranty *212*

9 Servicing Oven Motors   213

Fan motor problems *215*
Fan starts after 10 minutes *217*
The blower motor *218*
The stirrer motor *221*
The timer motor *225*
Convection timer motor *227*
Turntable motor assembly *228*
The vari-motor *232*
Auxiliary blower motor *234*
Motor problems *236*

10 Servicing Control Circuits   237

MCS 8100 basic oven operation *239*
Time and power level operation *240*
Temperature operation *241*
Memory operation *242*
The control board or DPC *243*
Control board circuits *244*
Panasonic key board checks *250*
Panasonic humidity control and sensor circuits *252*
Panasonic's NN-9807 MQ8897
    humidity sensor and DPC tests *253*
Panasonic oven temperature sensor thermistor tests *254*
Panel—no cook operation *254*
Panasonic's power control circuits *254*
Panasonic's other circuits *257*
Panasonic's variable power circuits *262*
Phase control circuit *262*
Multivibrator circuit and switching circuit *263*
Troubleshooting the rest *265*
Whirlpool preprogrammed test cycle *266*
Lightning damage *269*
Before you replace the control panel *270*
Installing a new control board *271*

11 Microwave Leakage Tests   273

Important safety precautions *273*
Required test equipment *275*
Radiation leakage monitor *275*
Holaday model H1-1500 survey meter *275*

Norelco rf radiation leakage tests *276*
Measuring precautions *277*
Where to check for leakage *277*
Tips on taking leakage tests *280*
Typical leakage tests *281*
Customer leakage tests *282*
Leakage record keeping *283*
Replacing the door *284*
Replacing waveguide assemblies *289*
Last minute checks *289*

12 Microwave Oven Case Histories   **291**

Nothing works *291*
Oven lights out *296*
Timer rotated but no operation *297*
Light on but no fan operation *298*
Flickering lights *300*
Cook light on but no turntable rotation *301*
Normal oven operation but no heat or cooking *303*
Oven goes into cooking cycle but then shuts down *311*
Erratic or intermittent cooking *313*
Oven lights stay on *316*
Keeps blowing fuses *316*
Hum or noisy operation *318*
Cannot turn oven off *319*
No fan or blower rotation *320*
Oven shock and leakage *321*
Too hot to handle *322*
Control panel problems *325*
Door problems *327*
Lightning damage *330*
Miscellaneous oven problems *330*

13 Important Dos and Don'ts   **335**

For safety *335*
For proper oven repairs *337*

14 Where and How to Obtain Parts   **341**

Service literature *345*
Warranty parts *345*
How to order parts *347*
Returning critical components *347*
Substitution of parts *348*
Service forms *350*

Appendix   Selected Circuit Diagrams   **353**

Glossary   **363**

Index   **367**

Acknowledgments

Many manufacturers have contributed microwave oven data and circuits for this book and to them I give a great deal of thanks:

Holaday Industries, Inc.

Matsushita Services Company

Narda Corporation

Norelco Service, Inc.

Without the help of many others this book would have been most difficult. A special thanks goes to Mr. Harry Foulds, National Service Manager of Matsushita Service Company (Panasonic), Mr. O.B. Walker, National Service Manager of Norelco Service Inc. (North American Phillips Corporation).

Introduction

*T*he printed circuit board was supposed to solve many wiring problems. The small transistor was designed to never break down. Like the microwave oven circuits, they were designed for simple and trustworthy operation—except that whenever you place mechanical and electronic components in a commercial unit, sooner or later you are going to encounter service problems. Who expected the microwave oven to be used as much as the color TV receiver? Today, those persons owning microwave ovens seemingly cannot do without them. The oven is used for preparing breakfast, lunch, dinner, and sometimes in-between snacks. The microwave oven has become an integral part of our new style of living.

Many ovens may be repaired by simply changing the ac line fuse. Often, the monitor switch may hang up, line voltage may overload, and nearby lightning may damage or open the fuse. Of course, even replacing the fuse requires removing the back cover. These oven fuses are chemical types, in ceramic sleeves, so one cannot peek inside to see if it is blown open. You should have a low-priced digital multimeter (DMM) to check for a defective fuse or poor interlock switch contacts.

Most anyone can repair microwave ovens with a few hand tools found around the house or upon the service bench and a DMM. Defective fuses, interlocks, motors, and components in the low-voltage circuits can be checked with a little knowledge and the digital multimeter. Never place hands or tools in the oven while operating. Remember, these low-voltage components operate from the ac power line. Leave the high-voltage (hv) system up to the professional electronic technicians. A Magnameter and leakage equipment should be used to locate the defective component in the hv circuits.

Before replacing or attempting to service the microwave oven, discharge the hv capacitor. This capacitor may hold a charge for several weeks in some ovens. Discharge the hv capacitor every time the oven is fired up. You may receive a shock of over 3,000 volts, which may injure or seriously damage a person. Discharge the hv capacitor with a pair of long insulated screwdrivers. Servicing microwave ovens can be extremely dangerous if you do not know what is going on.

The microwave oven has gone through several changes in the last five years. Most of the early ovens were manually operated. Today, practically all ovens are operated with push buttons or touch pads, like those on a calculator or computer. The new microwave ovens may be set for designated food—time is calculated, and by pushing a preheat and start pad, cooking begins and shuts off automatically for the selected time. Some ovens have variable-power cooking when touching the power level pad several times to select high, med-high, med, med-low, and low power cooking. Multiple sequence cooking may be programmed for several cooking sequences, switching from one variable power setting to another automatically. The electric brain provides a wide variety of cooking programs and all are controlled with a microprocessor in the control panel.

In this book, you will find practical and easy service methods to help locate defective components in the microwave oven. Various photos and drawings illustrate the many oven components. Although it is impossible to deal with every trouble that can possibly be found in the microwave oven, you will find the great majority of problems covered and resolved in the book.

Circuit diagrams, valuable service information, data, and drawings of the Norelco and Panasonic ovens are found scattered throughout the book. Fifteen different ovens with symptoms and method of repairing are found in over 200 case histories. Thirteen years of troubleshooting and repairing microwave ovens provides almost every trouble found within the microwave ovens. The microwave oven is here to stay. Now you too, can make these easy repairs discussed within this book.

How the low-voltage circuits operate with every kind of problem is found in Chapters 1 through 4. Chapters 6 and 7 discuss hv problems. The next chapter explains, in length, how to locate and replace the defective magnetron. The defective magnetron tube voltage and magnetron tests are easily explained. How to locate, remove, and install the control board circuits is discussed in Chapter 10. You can tell if the microwave oven is cooking as it should in Chapter 11. Over 200 actual microwave oven case histories are given in Chapter 12. Just reading this chapter may help solve the very problems you are now servicing. Some important dos and don'ts are listed in Chapter 13. Where do you find those important part replacements, after locating the defective component? Last, but not least, a list of major microwave oven manufacturers who can supply parts, and that most important oven circuit diagram finish up the last chapter.

Since the first edition went through the sixth printing, I have updated all chapters and included many different servicing data. New methods of servicing the control boards and new circuits are found throughout the book. Besides the added oven information, the book contains over 143 new photos and drawings. Remember the danger factor in repairing microwave ovens.

Chapter **1**

Some basic facts

Before long just about every home will probably have a microwave oven. Right now, microwave ovens are selling all over the USA and about every major appliance manufacturer is building them. Once you own a microwave oven, you cannot do without it, just as you cannot do without a TV. Although microwave ovens may not require the same amount of service as a TV set, they do break down like any other electronic/mechanical device (FIG. 1-1).

The microwave oven is an electrical and electronic appliance that is designed to warm or cook food within minutes. Power is applied to the magnetron tube after a sequence of buttons, switches, and other components are energized. The magnetron tube radiates rf energy that is tunneled to the oven and cooks the food placed in the oven cavity—food may be cooked within minutes or seconds.

DIFFERENT TYPES OF OVENS

Today, the microwave oven is found wherever food is prepared and served. A small compact oven is used in trailers, motor homes, and small apartments (FIG. 1-2). A standard tabletop microwave oven with 600 to 700 watts of power is found in most homes. Some of these tabletop ovens combine a separate ac heater-type coil for cooking and is called the convection oven. The commercial oven is found in restaurants, hospitals, and nursing homes.

MICROWAVE OVEN FREQUENCY

Microwaves are electromagnetic waves of energy. They are similar to light, radio, and heat waves. Microwaves have many of the same characteristics

I-I   Although microwave ovens may not require the same amount of service as the TV, they do break down as does any electrical/mechanical appliance.

I-2   Here are three different microwave ovens with different power ratings.

as light waves. They travel in a straight line, can be generated, transmitted, reflected, and absorbed. In a microwave oven, the magnetron tube generates the microwaves. They are transmitted to the oven cavity, reflected by the sides of the oven area, and then absorbed within the food that is in the oven cavity.

The microwave wavelength is relatively short compared to light. The frequency of a microwave oven is 2,450 million cycles with a wavelength of under 5 inches. There are three frequency bands alloted for microwave operation by the Federal Communication Commission. They are at 915 megahertz, 2,450 megahertz, and the highest one, 5,500 megahertz. Today, almost all of the microwave ovens operate at 2,450 megahertz.

The magnetron tube generates the cooking energy in a microwave oven. Materials with a high moisture content (like most foods) will absorb microwave energy. The food is made up of millions of molecules per cubic inch, which align themselves with the microwave energy when it enters the food at 2,450 megahertz. The microwaves are changing polarity every half cycle and are oscillating back and forth, creating friction. The friction between the molecules converts the microwave energy to heat and in turn cooks the food (FIG. 1-3).

1-3 The magnetron tube is the heart of the microwave oven. The magnetron gives off rf (radio-frequency) energy.

Microwaves can be reflected in the same manner as light. These short electromagnetic waves of rf energy pass through material such as glass, china, paper, and most plastic. Materials such as aluminum foil and stainless steel tend to reflect the microwaves while ordinary steel may absorb some microwave energy. Metal material placed in the oven cavity should be used only as recommended by the cooking instructions.

The sides of the oven cavity are constructed of metal so as to remain cool while the food is cooking. Although the sides and bottom of the oven may appear warm, this is caused by the transfer of heat from the food. The front door remains cool and it contains a metal plate with perforated holes. These holes reflect the microwaves but allow light to enter so you can see inside the oven cavity. Remember: water absorbs microwaves and begins to boil, while light passes through water.

Food prepared in a microwave oven should be cooked all the way through and not from the inside out. In some ovens, the food must be cooked for a few minutes, turned, and cooked again. You may find that other ovens have a device that rotates the food, providing for even cooking throughout. All food preparation and cooking should always follow the oven manufacturer's instructions.

FUNDAMENTAL FUNCTION OF THE MAGNETRON

The cathode is located in the center of the magnetron and is a filament that boils off electrons when it is hot. The cathode is connected to the negative side of the power supply, which has a potential of approximately 4,000 volts with respect to the anode, which is connected to the positive side.

The 4,000 volts potential is produced by means of the high voltage transformer and doubler action of capacitor and diode, of which a detailed circuit description is given later.

The electrons are negative charges, which means they are strongly repelled by the negative cathode and attracted to the positive anode.

The electrons would travel straight from the cathode to the anode if the 4,000 volts potential were the only force acting in the magnetron (FIG. 1-4). However, the magnetron is a type of diode with a magnetic field applied axially in the space between the cathode and anode by means of two permanent magnets (FIG. 1-5).

I-4 The electrons travel from cathode to anode (courtesy of Matsushita Electric Corporation of America).

If a magnetic field of sufficient strength is applied between the cathode and the anode, an electron would travel in a path almost at right angles to its previous direction, resulting in a circular motion of travel to the anode (FIG. 1-6). Eventually, it would reach the anode.

This circular motion by the electrons induces alternating current in the cavities of the anode. When an electron is approaching one of the seg-

1-5 Two permanent magnets are found between cathode and anode (courtesy of Matsushita Electric Corporation of America).

1-6 The electrons travel in a circular motion to the anode (courtesy of Matsushita Electric Corporation of America).

1-7 When the electron approaches one of the segments between two cavities, it induces a positive charge (courtesy of Matsushita Electric Corporation of America).

ments between two cavities, it induces a positive charge in the segment (FIG. 1-7). As the electron goes past and draws away, the positive charge is reduced while the electron is inducing a positive charge in the next segment. This inducing of alternate currents in the anode cavities can be thought of as a lumping together of the resonant circuit (FIG. 1-8).

1-8   Lumping the resonant circuits together (courtesy of Matsushita Electric Corporation of America).

In the actual operation of the magnetron, the electrons crowd together as they go around and, influenced by the forces of high voltage and the strong magnetic field, form a spoke wheel pattern (FIG. 1-9). This crowd of electrons, which has much stronger energy than a single electron, revolves around the anode and eventually reaches the cavities resulting in the continuous oscillation of the resonant circuits. The high

1-9   The strong magnetic field forms a spoked wheel pattern (courtesy of Matsushita Electric Corporation of America).

frequency energy, produced in the resonant circuit (cavities), is then taken out by the antenna and fed into the oven cavity through the wave-guide.

SOME DOS AND DON'TS FOR SAFETY

To prevent the risk of burns, fire, and electrical shock, the owner or operator should observe the following guidelines:

- Do follow the microwave oven instructions before attempting to cook. Don't forget the oven is a high-voltage and high-current appliance. Extreme care must be taken at all times. This does not mean you have to be afraid to operate the microwave oven.

- Do make sure the oven is installed properly. Most ovens include a three-prong plug. Pick one up at your local hardware store or electrical dealer (FIG. 1-10). Do not under any circumstance cut or remove the third ground prong from the power plug. The small flexible grounding wire with a spade lug will screw under the plate screw of the ac outlet. Use a voltmeter to determine if the center screw of the ac receptacle is grounded.

1-10 The microwave oven should be grounded through a three-prong ac plug. Do not cut off the third ground prong. Connect the flexible lead to the ac outlet plate screw.

- Do install a separate outlet from the fuse box for the microwave oven. Do not use an ordinary two-wire extension cord to operate the oven. Contact your local electrician or the appliance dealer who sold you the oven.

- Do make sure the ac power outlet is never under 105 volts or over 125 volts for proper oven operation. The microwave oven is designed to operate from 115 to 120 Vac. Don't just plug the oven into an overloaded outlet where several appliances are already tapped into it. If you do, the oven may be erratic in operation and never cook properly.

- Do install or locate the oven only in accordance with the installation instructions. Don't install the oven where the side or top opening may be blocked. The hot cooking air must escape and fresh air must be pulled into the oven for normal operation.

- Don't operate the microwave oven if the power cord or plug is damaged. Do repair the three-wire cord making sure the ground wire is intact. Check the continuity from the metal cabinet to the ground wire terminal with the low range of an ohmmeter.

- Do supervise children operating the oven. Most microwave oven problems are related to improper oven operation.

- Don't use the oven outdoors. Do not let the cord hang over edges of a table or the cooking counter. Keep the ac cord away from heated surfaces or a nearby cooking stove.

- Don't use metallic cooking containers. Use only cooking utensils or accessories of the type recommended by the manufacturer or microwave cookbooks. Some sealed containers in glass or plastic jars may build up pressure and explode.

- Don't use any type of material that may explode in the oven. A regular paper sack with food inside may explode and cause a fire if steam or air holes are not punched in the top of the sack. A fire inside the oven may cause plastic heat shields and front oven coverings to melt. This may result in a fairly expensive oven repair.

- Do keep the oven cavity area spotlessly clean. Fatty items such as bacon may in time collect grease behind plastic shelf supports on plastic microwave guide covers causing excessive arcing and damage. Use liquid window cleaner or a very mild detergent with a soft clean cloth over the face and interior surfaces. Do not use any kind of commercial oven cleaner inside the oven area. Use a paper towel or sponge to blot up and remove spills while the oven is still warm.

- Do pull the power cord of the oven when fire is discovered inside the oven cavity. Keep the oven door closed to help smother the fire.

- Do clean out the top area or exhaust areas for signs of food pulled up by the fan (FIG. 1-11). Have the serviceman brush out these areas when the oven is serviced. A collection of food particles may in time provide a persistent odor. Some odors may be eliminated by boiling a one cup solution consisting of several tablespoons of lemon juice dissolved in water in the oven cabinet.

- Don't change the operation of the oven while it is in operation. This is a good way to blow the fuse inside the microwave oven. Touch the stop button and start all over again.

I-II Keep the top area clean of food particles. The exhaust fan pulls the food up and should be cleaned out with a brush. The back cover must be removed to get at this area.

- Don't assume the microwave oven is defective if lines are streaking across the face of a TV set in a nearby room. Most ovens cause some type of interference to nearby TV receivers.
- When the microwave oven fails, make sure the three-prong cable plug is installed in the ac outlet. Make sure the outlet is alive by plugging in a lamp or radio. Do go over the operation instructions very carefully before calling the serviceman.
- Don't remove the back cover of the microwave oven unless you know what you are doing. Keep the power cord pulled when the back cover is off. **NOTE:** It's possible to receive a dangerous shock or possibly be killed in trying to just change the 15 amp fuse. The highly charged capacitor must first be discharged before touching any electrical or electronic component. (FIG. 1-12).
- Do not operate the oven with nothing in it.

OVEN INSTALLATION

Make sure the oven is located on a firm table base to prevent vibration. The oven should be installed with at least two or four inch clearance on

I-12 Don't try to change the 15 amp fuse until the hv capacitor is discharged. Use a couple of screwdrivers to short out across the capacitor terminals.

all sides of the oven cabinet. Of course, this depends upon where the oven vents are located. Leave at least one inch or more clearance at the top of the oven.

Try to avoid installing the oven near a heat source, heat duct, or cooking range. Keep the oven out of the sun where temperatures may exceed over 110 °F. The microwave oven will not operate if it is too hot. Avoid using the microwave oven where humidity is very high. Position the oven away from radios and TVs, for the microwave oven may interfere with the reception.

A microwave oven will operate successfully when adequate electric power is supplied to it. It's best to install a separate outlet for the oven. Since the oven is protected with a 15 amp fuse, additional appliances on the same circuit may overload the circuit and blow the house fuse.

All ovens should be grounded for personal safety. A three-prong plug is used to supply power and ground the metal oven cabinet. Make sure the third prong of the two-prong standard outlet is grounded. If installed by the appliance dealer, have the oven checked for proper grounds or have it checked by a local licensed electrician. While the electrician is present, have the power outlet checked for proper fuse protection and correct operation voltage (115 to 120 Vac).

SERVICE TECHNICIAN CAUTIONS AND WARNINGS

The microwave oven is a high-voltage and high-current piece of equipment and extreme care must be used while servicing it. Before taking off the back cover or outside wrap, remove your wristwatch. Make sure the oven is unplugged at all times when replacing and testing components.

1-13 Never stick a tool in or try to connect test equipment while the oven is running. Pull the power cords and discharge the hv capacitor before attempting to repair oven.

Never stick a tool or hand inside the oven while the oven is operating (FIG. 1-13).

Before checking any component or wiring in the oven, discharge the high-voltage capacitor just like you discharge the CRT anode connection before working around the high voltage section of a TV. The big difference between the high voltage of a TV receiver and a microwave oven is the amperage—you may be *severely shocked or killed* if the high-voltage capacitor is not discharged. Always discharge the capacitor with an insulated-handle screwdriver before working in the high-voltage area.

Make sure the oven is properly grounded before attempting to service the oven. If you're not certain, clip a flexible wire from a water pipe or fuse box to a metal screw on the back cover or metal chassis. Do not use a regular two wire extension cord to operate the oven while servicing.

The microwave oven should not be operated with the door open. Do not for any reason defeat the interlock switches at any time. Check and replace all defective monitors and latch switches when the oven will not shut off with the door open. The oven should never be operated if the door does not fit properly against the seal. Check for broken or damaged hinges. Visually inspect the seal gasket for possible cut or missing pieces. Check the gasket seal area for foreign matter. Make sure the oven door is mounted straight and fits snug against the oven. Readjust the door when play is felt between door and oven. Always, remember to check all interlock switch functions before the oven is returned to the customer.

While servicing a microwave oven, it's best to have the service bench away from the customer and not to talk to anyone while working on the oven. Outside distractions sometimes may cause you to make a mistake that may lead up to an injury. Always keep your hands out of the oven operation area when the oven is operating.

After making all repairs and service adjustments, check the oven for excessive radiation. To ensure the oven does not emit excessive radiation and to meet the Department of Health and Human Services guidelines, the oven must be checked for leakage with an approved radiation meter.

Especially check for radiation around the door and vent areas. If by chance the radiation leakage is greater than 5mW/cm² or someone has been hurt pertaining to microwave radiation leakage, report this to the oven manufacturer.

Although there are a few components that may be substituted or other manufacturer's components that may be used instead, most oven parts should be replaced with the original part number. These may be obtained from the manufacturer or oven distributors. When replacing control panel circuits, do not touch any part on the board as static electricity discharge may damage the control. Usually, these controls come packed in a static-free wrap and carton.

BASIC SERVICE PRECAUTIONS

When servicing any microwave oven there are several basic safety precautions to follow:

Microwave oven door

Do not try to operate the oven with the door open, an improper seal, and/or damaged door hinges. If the oven has been dropped, inspect the door, seal, and hinges. Make sure the door closes properly without any foreign matter inside the door.

Extension cords

Always, make sure the oven is grounded with a three-prong plug wall receptacle. Use only a three-wire extension cord between oven and wall outlet. It is best to have a competent electrician install a three-prong wall receptacle.

Unplug oven

Never reach into the oven component area while the oven is operating. When making ohmmeter or continuity tests, unplug the oven from the ac power line. If ac voltages are to be made, clip on the voltmeter with oven off. Do not hold the voltmeter test probes.

Shorting capacitor

Before checking any components in the oven, short out the two capacitor terminals. Always *discharge the capacitor each time* the oven is energized. Failure to discharge the capacitor may result in severe electrical shock.

Door interlock switches

Do not defeat or by-pass the interlock door switches. Each interlock should be checked for proper operation.

High-voltage section

The power transformer, capacitor, diode, and magnetron tube may contain high voltage (hv) when the oven is operating. When the oven shuts off, discharge the hv capacitor before attempting to do any type of service. Be careful when taking high voltages in the hv section.

Magnetron tube

Check the magnetron level and secured bolts to eliminate hot spots or radio frequency (rf) arcing. Always replace the rf gasket found around the magnetron tube antenna. Do not use metal tubes near the magnetron as it can jerk them out of your hands.

Microwave leakage tests

Always make a microwave leakage test with the approved radiation meter after all repairs are made. This should be done before oven front and side clean up procedures.

CHECKING THE OUTLET RECEPTACLE

The microwave oven should have a separate outlet and be protected with a 20 amp fuse. The outlet should be polarized and grounded. An improper ac outlet may provide electrical shock for the operator. Check the ac outlet for correct ac voltage (117−120 Vac) (FIG. 1-14A). Determine if the left side of the outlet is grounded (FIG. 1-14B). The right side of the outlet should be hot, if the ac receptacle is wired up correctly (FIG. 1-14C).

PULL THE PLUG, PULL THE PLUG

Always remember to pull the plug or cord before attempting to service the oven (FIG. 1-15). After each time the oven is fired up, pull the plug and discharge the hv capacitor before putting hands in the oven. Pull the plug before clipping test equipment to components in the oven. Pull the plug before taking continuity or resistance measurements of oven components. If, by accident, you are shocked by the hv capacitor, and end up with a metallic taste in your mouth, you know how careless and lucky you are.

LIGHTNING PROTECTION

Many of the new ovens have metal oxide varistors in the power line to protect the oven from lightning and power line surges (FIG. 1-16). These oxide varistors can be added to ovens that do not have this type of protection. Solder in the oxide varistors at the power cord fuse terminals inside the oven (FIG. 1-17). Now a heavy line voltage surge or lightning a block away will cause the varistor to arc over and blow the oven fuse. Check the house fuse if the varistor and main fuse are blown.

1-14A Checking for correct voltage.

1-14B Checking for good ground.

1-14C Checking for correct polarity.

1-15 Pull the ac plug or cord before attempting to service the microwave oven.

1-16 Add an oxide varistor to protect the microwave oven from power line surges and lightning.

1-17 Connect the varistor across the power line after the ac oven fuse.

BASIC OVEN COMPONENTS

Some microwave ovens have more or different basic oven components than others. Basically, all microwave ovens operate the same except some ovens have many additional features. Here is a list and brief description of most of the components found in a microwave oven.

Bleed resistor

The bleed or bleeder resistor is around 10 megohms and is found across the hv capacitor. When the oven is turned off, the bleed resistor discharges the capacitor within one minute (FIG. 1-18). Always discharge the hv capacitor before working inside the oven.

High voltage circuit

3.3 v AC

1-18   In some models, the bleed resistor discharges the hv capacitor when the oven is turned off.

Blower motor

The blower motor draws cool air through the blower intake directed at the magnetron tube. Since heat is developed within the magnetron tube, the tube is kept cool with the blower fan blades. Most of the hot air is exhausted directly through the vents of the back plate (FIG. 1-19).

Buzzer or speaker

The buzzer or speaker is used in ovens with the power control circuits to indicate when the cooking process is finished. The audio device is located on the control circuit board (FIG. 1-20).

Capacitor

The oil filled capacitor is located in the hv doubler circuit (FIG. 1-21). Ac voltage (2,500V) from the power transformer charges the capacitor

1-19 The blower motor draws cool air and keeps the magnetron cool during the cooking cycle. The oven may cut out after several minutes of operation if the fan is not rotating.

through a silicon diode increasing to approximately 4,000 peak volts. Always remember to *discharge the two capacitor terminals* before attempting to check or replace any component in the oven operating area.

Cook relay

The cook relay, when energized, closes the ac circuit so that the voltage is applied to the primary terminals of the power transformer. Sometimes this relay may be referred to as the power relay.

Cook switch

Usually the cook switch is the last component to be punched after the oven settings. The cook switch completes the circuit to the power trans-

1-20 The buzzer or piezo speaker indicates when the cooking process is finished.

1-21 The oil-filled capacitor is located within the hv doubler circuit. Locate the large capacitor and discharge the two top terminals before even changing the fuse.

former. A relay may be energized by the cook switch or power is completed through timer contacts, thermal cutout, and primary interlock switch.

Control circuit

The control circuit may consist of an electronic control board mounted behind the front control panel. In some ovens, the control circuit may be called the electronic controller. You may find both the front panel and the control circuit board are replaced as one unit (FIG. 1-22).

1-22 A control circuit may consist of the front panel control board. In some models, both the cover and the control circuit boards are replaced.

Defrost switch

The defrost switch is normally closed and allows the circuit to be completed for normal operation. The defrost circuit is made up of the defrost switch, defrost timer switch, defrost motor, and cam. Placing the defrosting lever in the on position opens the normally closed defrost switch and completes the circuit through the defrost timer switch.

Defrost timer switch

The defrost timer switch is opened and closed by the cam attached to the motor shaft. Actually, the defrost time switch is in parallel with the defrost switch (FIG. 1-23). The defrost and timer switches are similar to the vari-switch in other microwave ovens.

1-23 The defrost timer is in parallel with the defrost switch. This switch is opened and closed by a cam attached to the defrost motor shaft.

Door seal

The primary door seal is sometimes called a choke. A choke cavity reflects the microwaves back into the oven cavity. The choke cavity is filled with a material that is transparent to microwaves, called polypropylene.

Door latch

The door latch is an electrically operated mechanical device that prevents the door from being opened when the oven is in operation. The door may remain latched until the electrical circuit is interrupted. In some units, the door latch may be made of solid cast material or have a plastic spring-operated latch assembly.

Heating element

A separate heating element may be used for browning and cooking food in some microwave ovens. The heating element is located at the top of the oven cavity. The heating element may be controlled with the temperature control unit.

Interlock switches

You may find from two to five interlock switches in the various microwave ovens. Primarily, the interlock switches are activated by closing the oven door (FIG. 1-24). A cook cycle cannot begin until the door is closed. No microwave energy is emitted while the door is open.

Fuse

Most ovens are protected with a 15 amp chemically active fuse. Replace with the same type of fuse. You may find a fuse resistor circuit in other

Fuse

120 V
ac

Monitor
switch

Secondary
interlock
switch

1-24   You may find more than two interlock switches in the various microwave ovens. You may replace more interlock switches than any other component in the oven.

Glass resistor

Amana RR-40
model

Metal
fuse
assembly

1-25   A 15 amp chemical fuse protects the oven circuits from an overloaded condition. You may find a fuse resistor circuit in other ovens.

ovens (FIG. 1-25). The fuse resistor assembly senses an increase in current in the transformer secondary circuit and opens up the primary transformer circuit.

Fluorescent tube

Fluorescent display provides the time and operation numbers so they are visible to the eye. The fluorescent tube is controlled by a microcomputer or control IC (integrated circuit) located upon the control board.

Gas sensor

The gas sensor is a semi-conductor device that detects vapor, aroma, moisture, and humidity coming from the food as it cooks. The sensor found in a few microwave ovens adjusts the cooking time for various foods. This sensor is controlled by the control board (FIG. 1-26).

1-26 The gas sensor detects vapors, moisture, and humidity from the food in the oven.

Humidity sensor

The humidity sensor detects steam during the cooking period. In the Panasonic ovens, the sensor is located close to the exhaust vents where steam generated from the food is exhausted with circulating air (FIG. 1-27).

Light switch

The light switch turns the oven light on so the operator may view the contents in the oven cavity. This switch may turn the light on when the door is opened or closed. Usually, no oven light indicates a blown fuse, defective switch, or defective light. You may find more than one oven light in some ovens.

Magnetron

The magnetron is a large vacuum tube in which the electrons flow from the heated cathode to a cylindrical anode surrounded by a magnetic field (FIG. 1-28). These electrons are attracted to the positive anode, which has a high dc voltage. The magnetron tube oscillator has a very high frequency

1-27   The humidity sensor detects steam during the cooking cycle (courtesy of Matsushita Electric Corporation of America).

1-28   The magnetron is a large vacuum tube in which the flow of electrons is from the heater to the cathode to a cylindrical anode surrounded by a magnetic field. A high voltage is found on the anode with a low-voltage heater used to make the tube light up.

of 2,450 megahertz rf energy. The radiated rf energy cooks the food placed in the oven cavity.

Monitor or safety switches

The monitor switch is intended to prevent the oven from operating by blowing the fuse in case one of the other interlock switches fails to open when the door of the oven is opened (FIG. 1-29). This safety switch has normally open contacts when the door is closed. The entire ac line is placed across the monitor or safety switch when one of the other interlock switches fails. This blows the 15 amp fuse and prevents radiation from reaching the operator.

120V
ac

Monitor
switch

To control
circuits

1-29   The monitor switch protects the operator if, for some reason, the oven does not shut off when the door is opened. Change both monitor and interlock switch when one is found to be defective.

Oven lights

Most oven lights come on when the door is opened or when the oven begins to cook. The oven light may be located in the circuit after the triac device or control relay (FIG. 1-30). The oven light is located outside of the oven proper and shines through perforated holes.

Probe jack

The probe jack is located inside the oven and provides an easy method of plugging in the temperature probe. In some ovens, when the cooling probe is out of the jack, remember to turn the cooking control off or the oven will not operate. Always remove the temperature probe when the oven is empty. Do not let the probe touch any metal sides when in operation.

Rectifier

A special silicon diode is a solid-state device that allows current flow in one direction, but prevents current flow in the opposite direction. The

1-30   The oven light is found in the circuit after the control relay or triac assembly (courtesy of Matsushita Electric Corporation of America).

1-31   The silicon diode acts as a rectifier changing alternating current (ac) into pulsating dc. Replace the diode if it is running very warm.

rectifier and capacitor provide a hv doubler circuit. The diode acts as a rectifier by changing alternating ac current into pulsating dc (FIG. 1-31).

Relays

The cook or power relay provides ac voltage to the power transformer circuits. In some ovens, the coil of the cook relay may be energized by the control unit (FIG. 1-32). A surge current relay, used in conjunction with a

I-32 The cook or power relay provides ac voltage to the power transformer circuits. In some ovens, the relay may be energized by the control unit.

low value resistor, may be used as a current limiting device. The contacts of the select relay completes the circuit to the fan motor and oven light.

Resistors

A resistor is an electronic component that limits the flow of current or provides a voltage drop. The bleeder resistor across the hv capacitor lets voltage bleed off to chassis ground, through the hv diode. A fuse resistor in some ovens acts as a sensing device to an increase of current in the transformer secondary circuit and opens the transformer primary circuit. You may find a surge resistor located in the ac input circuit of some ovens to prevent damage to the oven circuits when power line outage or lightning occurs. A test resistor is used to measure the current of the magnetron tube.

Rotary antenna

Usually, the rotating antenna is located in a plastic tower in the middle of the cradle. The antenna assembly has several blades that rotate to stir the rf energy over the food in the oven cavity. The rotary antenna may be rotated with a separate motor or with a pulley upon the fan blower motor. In the early ovens, the moving antenna was found on top of the oven cavity. Now the antenna assembly may be found under the bottom floor shelf (FIG. 1-33).

Stirrer motor

A stirrer motor is located at the top of the oven cavity. The motor rotates or spreads the rf energy for even cooking in the oven (FIG. 1-34). Some ovens turn the food instead of using a stirrer motor. The stirrer motor may incorporate a separate ac motor or the fan may be rotated by a pulley-belt arrangement on the oven fan.

Stirrer shield

The stirrer shield is located at the top of the oven cavity covering the rf waveguide opening. This shield prevents food or other particles from

I-33 Wiring diagram. The rotary antenna may be found under the bottom floor shelf in the Panasonic MN9809 microwave oven (courtesy of Matsushita Electric Corporation of America).

going into the waveguide opening. The shield may need to be replaced when bumped, cracked, or broken by food being placed in the oven. The stirrer shield may be constructed from polypropylene material, which is transparent to microwave energy.

Stop switch

The stop switch is a closed contact in series with the start switch and power relay. When the stop switch is pressed, the circuit is broken and oven operations are stopped.

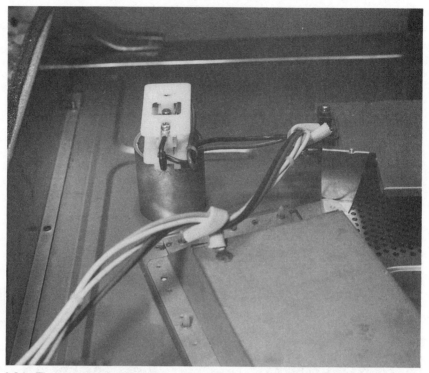

1-34 The stirrer motor or fan assembly is located at the top of the oven cavity. The fan blade circulates the rf energy for even cooking within the oven cavity.

Temperature fuse

The temperature fuse is in series with the ac oven fuse (20 amp). The thermal or temperature fuse may be found in some Canadian models. This fuse can be checked with the ohmmeter (FIG. 1-35).

Thermal cutout

The thermal cutout is designed to prevent damage to the magnetron tube if the magnetron becomes overheated. Under normal operation, the thermal cutout remains closed (FIG. 1-36). In some microwave ovens, an oven thermal cutout device is used in conjunction with the fan motor to cool the oven. The contacts of the oven thermal cutout are open at normal oven temperatures.

Thermistor

You may find a thermistor is used with the temperature control unit. The resistance of the thermistor in the temperature probe changes with the food temperature determining the time of cooking. The temperature in the oven cavity is detected through the resistance of the thermistor. The

I-35 The thermal temperature fuse is located in series with the 20 amp power-line fuse.

I-36 The magnetron thermal cutout is designed to prevent damage to the magnetron tube when the magnetron becomes overheated. Under normal conditions, the thermal cutout remains closed.

thermistor is a negative temperature coefficient component. A temperature probe is not found in all microwave ovens.

Thermostat

A thermostat may be found in convection type ovens. This thermal device senses the temperature of the convection air passing through the convection air guide. Different degrees of temperature may be selected by rotating the convection control shaft (90°−465 °F). Normally, the thermostat contacts are closed (FIG. 1-37). The thermostat is in series with the convection heating element.

Timer

You may find more than one timing device in the microwave oven. One timer may be used for from zero to five minutes of cooking time, while

1-37 The thermostat switch cuts the heating element out of the circuit when the oven reaches the temperature set by the operator.

1-38 The timer assembly turns on and off the amount of time allotted for cooking. A bell located on the timer is mechanically driven by the timer motor and rings once at the end of the cooking cycle.

the other timer has a longer cooking operation. The timer switch contacts are mechanically opened or closed by turning the dial knob located on the timer motor shaft (FIG. 1-38).

Timer bell

The bell striker is mechanically driven by the timer motor and rings once at the end of the cooking cycle, indicating that the food is ready. A defective timer assembly may prevent the bell from ringing.

Timer motor

The timer motor is energized through the timer contacts. When the timer reaches the 0 point on the scale, the timer switch opens the motor circuit and the cooking cycle stops. The rotating timer motor determines how long the oven remains in the cooking mode. In some ovens, the fan motor remains in operation until the contact of the oven thermal output opens up.

Transformer

The hv power transformer is a step-up voltage device operating from the 120 (Vac) power line (FIG. 1-39). The stepped-up voltage of the secondary winding is approximately 2,500 Vac and provides voltage to the voltage doubling circuit. A separate step-down secondary winding of the power transformer provides filament voltage to the magnetron tube.

I-39 The hv transformer has two separate secondary windings. One provides approximately 2,500 Vac to the voltage-doubling network, while the other is in the step-down filament winding.

Triac

You may find a triac module in various microwave ovens in place of a power relay. Usually, the triac module is controlled by the electronic controller circuit (FIG. 1-40). A defective triac may prevent the oven from going into the cooking cycle.

1-40 A defective triac may prevent the oven from going into the cooking cycle. In some microwave ovens, you may find a triac module instead of a power relay.

Turntable motor

In some ovens, a turntable device is used to rotate the food for even cooking. The turntable is rotated by a separate turntable motor assembly located at the bottom of the oven. A glass tray holds the food and rotates on several small rollers.

Vari-motor

The vari-motor assembly consists of a vari-motor, vari-switch, cam roller, gears, and mounting bracket. The motor rotates the vari-cam and activates the vari-switch on and off intermittently, supplying power to the power transformer. The repetition rate may be changed by turning the mode selector.

Vari-switch

The vari-switch is part of the vari-motor assembly and operated by the cam roller. If the variable cooking mode selector is set at warm, defrost, simmer, or roast position, ac power is supplied to the power transformer intermittently within a 30 second time base. Only a few microwave ovens contain a vari-motor and vari-switch assembly.

Varistor

You may find a varistor located in the ac input power circuit to protect the oven components from excessive power surges or lightning damage. The varistor will prevent the excessive voltage from entering these circuits. The varistor arcs over and opens the power fuse preventing damage to the oven circuits. These same type varistor components are found in the power line circuits of the latest TV receivers.

1-41   The magnetron tube is connected to the waveguide assembly. Rf energy from the magnetron antenna radiates power into the waveguide and the oven.

Waveguide

The magnetron tube is connected to the waveguide assembly. Rf energy from the magnetron tube's antenna radiates power into the waveguide and oven (FIG. 1-41). The waveguide channels the rf power from the magnetron into the oven. A waveguide cover, mounted at the top of the oven cavity, prevents food particles from entering the waveguide assembly from the oven area.

Like the TV receiver or any electronic device, various components may break down and cause many different symptoms within the microwave oven. With proper test equipment and tools, each defective component may be quickly replaced or repaired. Although most appliances and TV service centers have the basic tools for microwave oven repair, a few special added pieces of test equipment makes the task a lot easier and is really a must when servicing microwave ovens.

Chapter **2**

Required tools and test equipment

Only a few basic tools are needed to service minor problems with the microwave oven. Most of these hand tools are found around the house or in the average garage. A vom and vacuum-tube voltmeter (vtvm) is needed to check continuity and voltage in the various components. For accurate high voltage (hv) and current readings a commercial meter is rather handy when servicing many ovens. Most of the test equipment may be found in the ordinary TV shop. Additional test equipment and tools are described later in the chapter.

After each oven has been repaired or tested, a microwave leakage test *must be* made with a survey instrument. These tests are made to ensure the owner than no oven leakage is present to cause possible injury. Several leakage testers recommended by the industry are the Narda 8100, Narda 8200, Holaday 1500, and Simpson Model 380m (FIG. 2-1). They may be obtained through various electronic and electrical distributors.

PANASONIC'S MICROWAVE EXPOSURE WARNINGS

- Do not operate or allow the oven to be operated with the door open.
- Make the following safety checks on all ovens to be serviced before activating the magnetron or other microwave source, and make repairs as necessary:
    Interlock operation.
    Proper door closing.
    Seal and sealing surfaces (arcing, wean, and other damage).
    Damage to or loosening of hinges and latches.
    Evidence of dropping or abuse.

2-1 Here is a certified microwave leakage instrument—Simpson Model 380m. All leakage tests should be made perpendicular to the oven with a cone spacer.

- Before turning on microwave power for any service test or inspection within the microwave generating compartments, check the magnetron, waveguide, and cavity for proper alignment, integrity, and connections.

- Any defective or misadjusted components in the interlock, monitor, door seal, and transmission systems shall be repaired, replaced, or adjusted by procedures described in service manual before the oven is released to the owner.

- A microwave leakage check to verify compliance with the federal performance standard should be performed on each oven prior to release to the owner.

- Do not attempt to test the oven if the door glass is broken.

ATTACHING TEST EQUIPMENT

Most microwave oven manufacturers warn service technicians not to use ordinary voltmeters to take voltage tests in the ovens. The Magnameter is the only meter to use for voltage and current tests. Do not use a voltmeter unless it will test over 4,000 Vdc. Do not try to measure the filament voltage with a small voltmeter as the filament winding is at the highest dc voltage potential (FIG. 2-2).

2-2 Do not try to measure the filament voltage (3.3 Vac) with a small voltmeter because the winding is connected to the highest dc voltage source (courtesy of Matsushita Electric Corporation of America).

When making voltage tests, clip all meter test leads into the circuit. If temporary parts are to be added or substituted, do not hold them in your hands. Clip parts into the circuit with clip-on test leads. Keep your hands out of the oven when operating. Before opening up the microwave oven, remove your wristwatch whenever working close to or replacing the magnetron. People wearing pacemakers must consult their physician before attempting to service microwave ovens. It's wise to place an 8 ounce paper cup of water in the oven during testing operations.

BASIC HAND TOOLS

Although the average service technician may have the basic hand tools for oven repair, a list and description are given below:

Screwdrivers

Every home and shop has a lot of screwdrivers. You will need two different size Phillips screwdrivers, one large and one small. The larger Phillips screwdriver is needed to remove those stubborn screws that hold on the back cover. Often the nickel-plated metal screws are tightened with a power screwdriver and sometimes are difficult to remove.

Two regular size screwdrivers come in handy to remove the various components, besides prying up a stubborn component from the metal chassis. Select two long insulated type screwdrivers for discharging the hv capacitor. The two screwdriver blades are jammed against the capacitor

2-3 Various sizes of Phillips and straight screwdrivers are needed to work on a micro-wave oven. Choose a couple of long ones to discharge the hv capacitor.

terminals and sandwiched together to discharge it. A couple of small short screwdrivers come in handy when working in tight corners to remove or tighten those difficult to get at metal screws (FIG. 2-3).

Some ovens require a special Torx or LHSTix type screwdriver for removing power cords and other components.

A nut-driver set may speed up the removal and replacement of the various components in the oven. Besides a nut driver selection, a set of socket wrenches may come in handy to remove those large power trans-formers and the turntable motor assembly. A small crescent wrench will do if a set of sockets are not available. To remove some of those front knobs and shafts, keep a set of Allen wrenches handy.

Of course, a pair of regular pliers can be used in many ways around the microwave oven. The long nose pliers are handy when soldering cables or wires to the various parts. Select a good pair of long nose and side cutter pliers to make connections and remove wiring and cables from the replaced components.

Choose a regular 100 watt soldering iron or 150 watt soldering gun for most terminal connections. The larger cables and wires require more heat than those found in the TV chassis. You may find most connecting wires are crimped in ovens, but many have to be soldered. A small low-wattage iron is handy when working around the control boards.

Homemade tools

Select three or four sets of alligator clip leads or make them yourself. They may be purchased at electronic supply houses. Simply solder an alli-gator clip to a 12 inch flexible cable lead. The alligator clip ends should be of the rubber insulated type so they will not short out components when

Flexible test cable

2-4 Choose several alligator clip leads for temporary connections. They come in handy to short around a suspected interlock switch.

in tight places. These cables come in handy to short around interlock switches or to temporarily clip components in the circuit (FIG. 2-4).

Take a long bladed screwdriver, preferably a discarded Phillips type and solder a two foot cable with a large alligator clip at one end. A technician may use this tool to short the hv capacitor to chassis ground (FIG. 2-5).

Alligator clip

Long screwdriver

2-5 A two-foot cable, alligator clips, and a screwdriver are used to discharge the hv capacitor to common chassis ground.

Clip the alligator clip to a good ground and place the blade of the screwdriver against both capacitor terminals. The grounding tool is also used to ground out any voltage at the filament terminals of the magnetron tube, before removing the defective tube.

Another useful homemade tool is a fuse puller. Take a discarded screwdriver with a long stem and bend a curl, a u-shaped end so the tool

Bend hook

Old screwdriver

2-6 You can make your own fuse puller from a discarded screwdriver. Form a clip on the end to pull and pry out the 15 amp fuses.

will slip under the fuse (FIG. 2-6). Apply pressure against the fuse holder or pull up on the fuse at one end. Now the fuse can be removed. Sometimes the chemical fuses are located back inside the oven wall or mounted on the monitor switch assembly and may be difficult to remove.

Take a TV cheater cord and cut off the interlock portion to make a temporary ac connecting cable. Solder two regular insulated alligator clips to each end for easy attachment. This cable can be used to check the ac cable or to apply power directly to the power transformer leads to isolate the control circuit from the transformer circuits.

Pick up a pigtail light bulb socket for continuity and fuse blowing tests. These can be located in electrical departments or at your local electrical store (FIG. 2-7). Screw in a 100 watt bulb. Instead of blowing a chemical fuse each time a short appears in the oven or an interlock switch hangs up, the light will become bright with the full power line voltage across the pigtail test light. This troubleshooting method may save a lot of blown 15 amp fuses.

2-7 A handy test light is constructed from a pigtail light socket and alligator clips. These sockets can be picked up at most hardware or electrical stores.

VOLT-OHMMETER

The small pocket type vom is a handy tester in checking continuity of a suspected fuse, solenoid winding, interlock switch contacts, and also for tracing wiring cables. You should have one of these in the tool box when checking ovens in the home (FIG. 2-8). A low ohm range from R×1 or R×1k is sufficient. All continuity measurements should be made after the power cord is pulled and the hv capacitor is discharged.

2-8 A good continuity tester is a small vom found in most TV and work shops. These are handy to locate dead fuses and open components.

Only use the ac voltage range to check for ac power-line voltage around the microwave oven. Do not use the dc range in the high-voltage circuits. These voltages are above 2,000 volts and will quickly damage or ruin the meter. Only use the small vom for low ohm and power line ac voltage measurements.

DIGITAL VOM

The small digital vom comes in handy when taking very low-ohm measurements of transformer windings and filament continuity of the magnetron tube. Although the digital vom is not a required test instrument, accurate low-ohm resistance tests (less than 1 ohm) may save a lot of valuable service time (FIG. 2-9). A low-ohm measurement is necessary since the primary winding of the power transformer may be less than 1 ohm and

2-9 Although the DMM is not a required test tool for oven repair, it will accurately take resistance measurements of solenoids, magnetron heaters, and transformer windings, as well as fuses.

the secondary filament winding from 0 to 1 ohm. The resistance measurement of the filament or heaters of the magnetron tube may be from 0 to 1 ohm.

If a diode check is provided by the digital meter, this measurement may be used in checking diodes and transistors on the control board. Some technicians and oven manufacturers prefer to replace the control board instead of servicing it. The diode check is not accurate when testing the hv diode in the magnetron tube circuit. The milliampere scale may be used when the meter reads over 300 milliamps and a grounding test resistor is located in the hv diode circuit. Primarily, use the digital vom for continuity and low ohmmeter measurements.

DMM DIODE TESTS

The digital multimeter (DMM) is ideal to take low ohm-continuity tests on resistors, diodes, transistors, and IC components upon the control board. A good diode shows a measurement in one direction only (FIG. 2-10). The low ohm measurement in both directions indicates a leaky diode. Some technicians and oven manufacturers prefer to replace the control board instead of servicing it. The diode check is not accurate when testing the hv diode in the magnetron circuit with a low voltage powered DMM. Use a DMM with a 6 or 9 volt battery to provide any type of indication with the hv diode/rectifier.

2-10 DMM diode tests.

DMM TRANSISTOR TESTS

Transistors junction tests can be made with the DMM that has a diode test. Always place the red probes to the base terminal of an NPN transistor and the black probe to the base with an PNP type (FIG. 2-11). You may find one or two elements of the transistor open or leaky. A normal transistor may show a measurement from base to emitter and base to the collector terminals. A low measurement in both directions indicates a leaky transistor. Check for improper voltages and low ohm measurements to ground on each IC terminal to locate leaky or shorted IC components.

VTVM

Most TV shops have a vtvm for radio and TV servicing. The vtvm may be used to check for continuity and make voltage measurements in the microwave oven. The various resistance and voltage ranges are ideal to check the hv transformer, relays, magnetron tube, and to make voltage measurements on the control board.

The vtvm may be used to check the hv circuits of the magnetron tube. A special TV hv probe must be used with the vtvm for these measurements. Do not try to make hv measurements without the probe. If the vtvm has a polarity switch for voltage measurements, it makes an ideal hv instrument.

Remember, the high voltage around the magnetron is negative in respect to the oven chassis. Use the negative dc voltage range. When tak-

2-11 Transistor test with DMM.

ing hv measurements with the vtvm, keep the instrument off the oven, unless insulated from the metal cabinet. A piece of masonite, a book, or a piece of plastic will do.

THE MAGNAMETER

One of the latest voltage and current meters, called the Magnameter (model 20-226), is available from GC Electronics of Rockford, Illinois. The Magnameter is a specialized instrument used to speed up and simplify microwave oven testing. You may quickly take voltage and current measurements without possible danger of hv shock and injury of the magnetron circuits (FIG. 2-12).

The meter enables the technician to make both hv and plate-current measurements of the magnetron and hv circuit with one setup. You don't have to handle or change test leads during the voltage and current checks.

2-12 The Magnameter is an accurate tester to indicate a defective magnetron or hv circuits. Safe current and voltage tests upon the magnetron tube indicate a shorted, open, or slow-cooking magnetron tube.

The meter hookup is identical with positive and negative ground ovens. Most domestic ovens have a positive ground in the hv circuit.

The meter is well insulated with 10kV rated test-lead wires. A high impact plastic meter case prevents shock to the operator. No metal knobs or switches are exposed preventing possible shock. The meter can be held in your hands or set on top of the metal oven while taking critical voltage and current readings. The meter case is constructed of extremely tough thermoplastic material that resists breakage from all but the most severe abuse.

Besides the meter indication, a neon-warning light flashes when high voltage is present—whether the switch is in high (hv) or in low (current) position. Both readings may be obtained with just flipping the hi-low toggle switch. First, take the hv reading and then a current measurement.

The meter hand is rated from 0 to 10kV. You may consult the oven manufacturer's service information for acceptable hv limits. Most domestic ovens operate between 1.5 and 3kV, while commercial ovens operate from 2 to 4kV applied to the magnetron tube. The toggle switch must be flipped to high to measure the high voltage in kilowatts.

Flip the toggle switch to low for direct magnetron plate current measurement. In some microwave ovens (especially the early ones), a 10 ohm plate resistor is found between the hv diode and chassis ground (FIG. 2-13). Using this 10 ohm resistor is a quick method to take current readings without possible shock to the technician.

When no test resistor is found in the magnetron circuits, simply

2-13 Connect the 10 ohm test resistor between the hv diode and the chassis. Remove the positive end of the diode and clip the resistor to the diode and chassis ground.

remove the ground side of the silicon diode (+) and connect the 10 ohm test resistor, provided with the meter, between diode and chassis ground. Actually, you are inserting the test resistor in series with the hv diode and common ground terminal. Connect the green meter lead to the 10 ohm resistor for current measurement. Now, you simply multiply the meter reading × 100 to get the actual current reading of the magnetron tube. For instance, if the meter hand was at 2 then × 100 equals 200 mills of current. The current of most domestic microwave ovens will vary from 160 to 400 milliamperes while the commercial ovens may have a plate current from 200 to 750 milliamperes. Very seldom does the manufacturer include the operating current of the microwave oven in the service literature. It's wise to take a voltage and current reading on every oven and mark them on the service literature for future reference.

Connecting the meter

After removing the back cover of the microwave oven, *discharge the hv capacitor*. Double check—make sure the power plug is disconnected. Connect the black lead of the meter to common chassis ground. These insulated alligator clips are strong and will stay in place. Clip the red lead to the hv capacitor or hv diode of the hv side (FIG. 2-14). If in doubt, trace the hv capacitor lead going to the filament circuit of the magnetron tube.

Check for a 10 ohm 10 watt test resistor. If the hv silicon rectifier goes directly to chassis ground, remove the positive end of the diode. Usually this lead is grounded with a metal screw or nut. Now, insert the insulated 10 ohm test resistor provided and clip between the positive end of diode (end just removed) and common ground (FIG. 2-15). Most of the new ovens do not have a test resistor mounted in the hv circuit.

Double check the meter connection. Make sure each clip is snug and tight. Each cable end has a colored rubber alligator clip for easy identification (the red lead goes to the high-voltage side, the black lead to ground, and the green lead to the top side of the hv diode). The meter should be connected as shown in FIG. 2-16.

2-14 Connect the meter to the hv capacitor and diode. Clip the red lead to the negative side of the diode. Connect the black lead to ground.

2-15 When a 10 ohm resistor is not found in the hv circuit, remove the positive end of diode from the chassis ground. Connect the 10 ohm test resistor between the diode and ground. The current test is taken across the 10 ohm resistor.

2-16 Connect the Magnameter as shown in the schematic. Check the color code of each test clip for correct identification.

Taking voltage and current measurements

After the cables from the meter have been attached to the oven circuits, start the cooking cycle. Set the meter toggle switch to high. Plug in the power cord. Place a liter of cool water in the oven cavity. Turn the oven to maximum cook or heat and push the cook button. Now, read the high-voltage shown on the meter.

Change the toggle switch to the low position. Read the low voltage which is × 100 and this will give you the milliampere current reading. From these two readings you can quickly determine if the hv circuits or the magnetron tube are defective. Now, return the toggle switch to the high position. Pull the power cord. Discharge the hv capacitor with the discharge switch on the side of the Magnameter. The meter hand should return to zero and the warning light should be out, indicating the hv capacitor is discharged. To be sure, you may want to discharge the capacitor with two screwdrivers. Disconnect the meter and proceed to service the oven.

When no high voltage is seen on the meter in the high position, suspect a leaky diode, hv capacitor, transformer, or no ac voltage is supplied to the transformer primary winding (FIG. 2-17). You may find a grounded magnetron tube with no high voltage and no current reading. A shorted transformer and leaky diode may produce a low high-voltage reading (often below 1kV).

Excessive high voltage reading may be caused by a defective magnetron, open filament, open filament winding, or high-voltage wire. You

2-17 When no hv is measured, suspect a leaky diode, hv capacitor, or transformer. A leaky or shorted magnetron tube may show low or very little high voltage.

may have a low or no current reading with an excessive high-voltage reading from 3 to 5kV. A leaky or shorted magnetron tube may produce a high-current measurement from 350 to 600 mills (FIG. 2-18).

The Magnameter will quickly indicate problems in the hv circuit or outside the magnetron tube circuit. Erratic meter readings will indicate an intermittent magnetron tube or poor filament connections. Low current readings may indicate a weak magnetron tube that can cause the problem of taking too long to cook foods in the oven. A meter of this type may be a great addition to your present vom or vtvm.

2-18 Suspect a defective magnetron tube, open filament, open transformer winding, or hv wire with excessive high voltage. Check for a very low current reading with excessive high voltage.

COOKING TEST EQUIPMENT

The power output of the magnetron may be measured by performing a water temperature rise test. A cooking water test should be made after all repairs are finished. The cooking test may indicate a defective magnetron or hv component. When the oven takes too long to cook, suspect a faulty magnetron or related components and make the cooking test.

For making the cooking test, you should have a 600cc glass beaker or equivalent, glass thermometer 100 °C or 212 °F, and two 1 liter beakers or equivalent.

You may find that each oven manufacturer has a somewhat different water test (each oven test procedure uses water, a container to hold the water, a thermometer, and a stop watch). To ensure accurate water tests, the power-line voltage should be 115 to 120 Vac.

Here are several different manufacturers' water test procedures:

Panasonic NN-9807 and MQ8897BW cooking tests   The power output of the magnetron can be determined by performing a simple water heating test as described below. Necessary equipment—two 1 liter beakers, glass thermometer, and wristwatch or stop watch. Check the line voltage under load. Low voltage will lower the magnetron output. Take the temperature readings and heating time as accurate as possible.

1. Fill each beaker with exactly one liter of tap water. Stir the water using the thermometer and record each beaker's temperature (recorded as T1A and T1B).

2. Place both beakers on the center of cook plate. Set the oven for high power and heat it exactly for two minutes.

3. Stir the water again and read the temperature of each beaker (recorded as T2A and T2B).

4. Obtain the temperature rise by using the following formula:

Average temperature rise =

$$\frac{(T2A + T2B) - (T1A + T1B)}{2}$$

The normal temperature rise for these models should be 16° to 20 °F (8.9° to 10 °C) at the high power selection with the oven operating at the specified line voltage.

5. To obtain the power output, multiply the average temperature rise by 39 (or 70 if thermometer is in Celsius). Example: If you find average temperature rise of 18 °F, the oven output is 18 × 39 = 702 watts.

Oven A

1. Pour 1,000ml of cool tap water into a large beaker and stir with the thermometer to measure the water temperature. Make a note

of the initial temperature (T1). This temperature should be somewhere between 17 °C and 27 °C.

2. Place the beaker of water in the center of the oven and heat for 62 seconds at high power. Use the watch second hand instead of the oven timer.

3. When the time is up, stir the water with the thermometer and again measure the temperature. Record the final temperature (T2).

4. The output can be calculated with the following formula: wattage output (W) = (T2 − T1) × 70 (Celsius).

5. The maximum temperature rise is 10 °C—a minimum temperature rise is 8 °C.

Oven B

1. Fill the measuring cup with 16 ounces (453cc) of tap water. Measure the temperature of the water with the thermometer. Stir the temperature probe through the water until the temperature stabilizes. Record the water temperature.

2. Place the cup of water in the oven and set the cooking mode to full power. Allow the water to cook for 60 seconds while checking with a stop watch, the second hand of a watch, or the digital readout countdown.

3. Remove the cup and measure the temperature. Stir the temperature probe through the water until the maximum temperature is recorded.

4. Subtract the cold water temperature from the hot water temperature. The normal result should be a 10° to 36 °F (10.6° to 20 °C) rise in temperature.

Oven C

1. Select a 1 pint (2 cup) pyrex measuring cup and a thermometer with a range of over 180 °F.

2. Fill the pyrex container with one pint of tap water. Use the thermometer and record the water temperature.

3. Place the filled container in the center of the oven cavity and set the timer at 2½ to 3 minutes.

4. Time the oven for exactly two minutes with a watch or clock with a second hand.

5. Remove the pyrex container and stir the water with the thermometer to check the rise in temperature. If the oven is operating successfully, the temperature rise should be approximately 55°.

You should follow each manufacturer's cooking water test when working on a particular type of oven. All water tests indicate how the magnetron is functioning. If the water temperatures are accurately measured and tested for the required time period, the test results will indicate if the magnetron tube has low power output (a low rise in water temperature) that would extend the cooking time or high power output (a high rise in water temperature), which would reduce cooking time.

One microwave oven technician I know has devised his own cooking water tests. Three new and different microwave ovens were used in the test. Each oven was checked against a beaker filled to a certain level and marked at the boiling point (FIG. 2-19). The time of each oven was recorded at 3 minutes. When the water began to boil the time was recorded for each oven. A high and low line were drawn on the beaker. Most ovens with the same water level had very close water boiling times (2$^{1}/_{2}$ to 3 minutes).

2-19   Three different cups marked for power and water level. Each oven was recorded at three minutes.

When an oven takes longer than three minutes for water to start to boil, the technician knows trouble exists in the magnetron and hv section. In case the boiling point is more than 4 or 5 minutes, a voltage and current test of the magnetron tube will indicate a defective tube (FIG. 2-20). Of course, water boils at different temperatures in different parts of the country (depending on altitude) and this must be taken into account with this type of method. If the water boils below the 2 minute mark, he knows the magnetron tube is defective and usually is running red hot.

2-20 When it takes the oven longer to make the test water boil, check the current of the magnetron tube. Poor electron emission and filament problems may take the magnetron longer to heat.

If by chance you are on a house call to check a microwave oven and you forget to take a beaker or pyrex cup along, try a styrofoam cup. Today, most kitchens have such cups available. These cups do not melt down, although they do run warm when the water begins to boil. Fill the cup with tap water and take the three minute boiling test. You should know how full to fill the foam cup. After several house calls or cooking tests, you know whether the oven is cooking accurately or if it is defective. If in doubt, bring the oven in and apply the manufacturer's water cooking test.

In case the cooking test takes a longer cooking time or the customer complains of burning up the food, check the magnetron tube. A magnetron tube pulling heavy current (400 to 600 mills) indicates a leaky or shorted tube. When the oven takes too long to cook, check for very low current readings and suspect a weak magnetron. Replace the magnetron if the correct hv voltage is present but there is an improper current measurement.

LEAKAGE TESTS

Every microwave oven *must have* a leakage test taken after servicing the unit. All oven manufacturers require leakage procedures, besides most customers demand a leakage test. You may be asked to come to the home and only make a leakage test, since most people are scared to death of radiation. It's better to be safe than sorry—so take a leakage test for your own safety too.

There are several leakage test instruments recommended by the industry, Narda 8100, Narda 8200, Holaday 1500, and the Simpson Model 380m (shown in FIG. 2-21). The price of a leakage tester may run close to

2-21   Here is the Simpson Model 280m radiation leakage instrument. The instrument is used to check for rf radiation around the door and the oven outlets for possible leakage.

.$500. Two different low-priced leakage testers are discussed at the end of this chapter.

The purpose of any radiation instrument is to check the radiation leakage around the microwave oven door, outer panels, vents, and door viewing window. After replacing a magnetron tube, check for leakage around the magnetron and waveguide assembly. Be careful—avoid contacting any hv component with the radiation meter (FIG. 2-22).

2-22   Move the probe around the shaded area, the exhaust openings and air inlet openings (courtesy of Matsushita Electric Corporation of America).

The U.S. government has established a maximum of 5mW/cm² leakage while in the customer's home. Many oven manufacturers request that the leakage should never be more than 2mW/cm² (1mW/cm² for Canada).

The power density of the microwave radiation emitted by a microwave oven shall not exceed 1 milliwatt per square centimeter at any point, 5 centimeters or more from the external surface of the oven, measured prior to acquisition by a purchase, and thereafter, 5 milliwatts per square centimeter at any point 5 centimeters or more from the extreme surface of the oven.

THE NARDA 8100 RADIATION MONITOR

The Narda instrument measures radiation leakage in milliwatts per square centimeter (mW/cm²). A water load of 275cc (approx. 1¹/₃ cups of water) is placed in the oven and used as a load during leakage tests. The measuring probe should be used with a two inch cone spacer.

This tester may be operated on the internal rechargeable battery or from the power line. The battery may be charged from the power line. Switch the instrument to the 2,450 MHz position. A fast and slow meter response switch should be set in the fast position. This tester has an available alarm system warning the operator when high amounts of radiation may damage the meter. Set the alarm control to 50 which sounds the alarm when the meter reads 50% of full-scale deflection.

On ovens with an unknown leakage, use the high scale first and then switch to the low scale for low leakage. A test switch is used to check the battery and probe. The meter needle will not read above the minimum test mark on the meter if either scale is faulty. Set the meter to 0 with the 0 control.

Always check the battery and probe with the test switch before attempting to measure radiation leakage. Check both the battery and the probe test switch. Plug in the ac cord if the battery reading does not come up to the minimum test setting. Do not use the probe if the probe tests fail. The audio alarm will come on, during operation, if the probe becomes inoperative or disconnected.

THE SIMPSON 380 SERIES MICROWAVE LEAKAGE TESTER

The Simpson 380m is a portable, direct reading instrument designed for accurately measuring the amount of microwave leakage radiated by ovens, heaters, chargers, and other industrial equipment generating high power at microwave frequencies of 2,450 MHz. The nonpolarized, hand-held, wide range probe permits selection of five distinct ranges for power density measurements. This instrument has a slide switch, enabling the technician or operator to select the correct response time.

The Simpson 380m comes with a thermocouple probe, cone spacer, two 9 volt batteries, carrying case, and operating manual. When a new probe is required, the complete instrument must be returned to the factory for calibration. Probes are not interchangeable. Most radiation test equipment should be returned to the factory or factory authorized service centers once a year for overall check and calibration.

A six-position rotary-range switch is used to activate the tester, as well as to select the range for the radiation to be measured. The range selected is indicated by the marker. With the range switch set to either 2.5 or $2mW/cm^2$ position the lower (0 to 2.5) dial arc is read. The (0 to 10) dial arc is read when the range switch is set to the 10 or $100mW/cm^2$ position. Switch the range switch to the battery test position to check the condition of the batteries.

If the fast-slow switch (normally set to the fast position) that selects the response time of the meter remains at the set point (does not move towards zero) the probe is defective. Carefully read the manufacturer's operating manual before making any leakage tests. Make sure the batteries are good before making leakage tests. You quickly get the hang of things after three or four leakage tests.

You may contact the following companies for more information about their microwave leakage detectors.

HOLADAY INDUSTRIES, INC.
14825 Martin Drive
Eden Prairie, MN 55344

NARDA MICROWAVE, INC.
Plainview, NY 11803

SIMPSON ELECTRIC CO.
Elgin, IL 60120

PANASONIC'S CONFIRM AFTER REPAIR

After repair or replacement of parts, make sure that the screws of the oven, etc., are neither loose nor missing. Microwaves might leak if screws are not properly tightened.

Make sure that all electrical connections are tight before inserting the plug into the wall outlet. Check for microwave energy leakage.

Chapter 3

Microwave oven circuits

Today there are four basic and separate circuits in the microwave ovens. The low-voltage circuit consists of all components found operating from the power line. The high-voltage (hv) components are located from the hv power transformer to the magnetron circuits. All microwave ovens have the above circuits. Within the low voltage stages you may find several different circuits. The first ovens were manually operated. Then, automatic push-button cooking operated with an electronic control panel. The convection oven may operate with either a manual or electronic control oven.

DIFFERENT TYPES OF FUSES

Some microwave oven manufacturers use the common household fuse to protect the oven circuits. Many manufacturers provide fuse protection with a long chemical type fuse (FIG. 3-1). You may find another temperature type fuse in series with the ac oven fuse. Usually, the oven fuse is located in the power-line circuit before any switch operations.

Pull the power cord from the receptacle before replacing or testing the oven fuse. If the fuse is buried under wires and components, discharge the hv capacitor before removing old fuse. Replace the monitor interlock switch when the door is opened and the fuse blows at once. The fuse may open with overloaded low or high voltage components. The oven fuse is located in the low voltage circuits (FIG. 3-2).

LOW-VOLTAGE CIRCUITS

All components from the power-line cord to the primary winding of the hv transformer may be included in the low-voltage circuit. Each compo-

3-1 You may find a common 20 amp fuse or a long chemical fuse in the microwave oven.

3-2 The oven fuse is located in the power line of the low voltage circuits.

nent must operate to provide power-line voltage (117 to 120 Vac) to the transformer (FIG. 3-3). The failure of the fuse, interlock switches, timer, cook switch, controller, triac, and oven relays may prevent the oven from

3-3 In the low voltage circuits, each component must function to provide power-line voltage to the primary winding of the hv transformer. The ac voltmeter or a 100 watt bulb may be used to check the condition of the oven components.

operating. Other components in the low-voltage circuit may or may not prevent the oven from operating, but provide convenience of operation signals and indicators when troubleshooting.

Seventy-five percent of microwave oven problems are caused by open fuses and defective interlock switches. These two components must be operating correctly or no power-line voltage is applied to the primary winding of the hv transformer. You may find from three to six interlock or latch switches in the low-voltage circuits of microwave ovens. Besides providing power to the oven circuits, the interlock switches also provide safe oven operation.

Such low-voltage components as the blower, timer, turntable, and vari-motors may be used for indication when checking the low-voltage circuits. When a motor is not rotating, no ac voltage may be found at this point in the circuit or the motor may be defective. A dead oven light may quickly indicate insufficient power is getting to the oven circuits. When the cook light or controller display does not light up you may determine where the power-line voltage stops in the low-voltage circuits.

How an alternating electrostatic field is generated

The generation of an alternating electrostatic field can be explained by studying the generation of alternating current, which is widely known as the household power supply source. The alternating current (ac) takes place in a resonant circuit, which consists of a coil of wire and a capacitor. The coil and capacitor store energy respectively, but in different ways. When they are connected together, and connected to a source of energy, an alternating current is generated. We will see how this happens in the following sections.

How the coil stores energy in a magnetic field

A coil of wire carrying an electric current will generate a magnetic field around the coil. This field will have a north pole and a south pole at opposite ends of the coil exactly like the permanent magnet does. If the direction of current through the coil is reversed, the direction of the magnetic field also will be reversed. If the voltage source supplying the electric current is cut off, the magnetic field around the coil collapses. This collapsing field generates a voltage in the coil which for a short time keeps the current flowing in the same direction. In this way, the energy stored in the magnetic field is returned to the circuit. This ability of coil to store energy is called an "inductance" (FIG. 3-4).

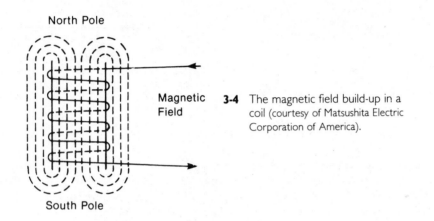

3-4 The magnetic field build-up in a coil (courtesy of Matsushita Electric Corporation of America).

How the capacitor stores electric charges

An ordinary capacitor is made of two metallic plates, close together but separated by air, paper, oil, mica, or other nonconducting material. If the two plates are connected to the terminals of a battery or other source of energy, the plates of the capacitor will be charged, with negative electrons on one plate and positive charge on the other plate (FIG. 3-5).

3-5 The charges built up on the plates of a capacitor (courtesy of Matsushita Electric Corporation of America).

When the connection to the battery or other voltage source is reversed, the electrons will travel through the external circuit until the polarity of the charges on the capacitor is reversed. This current through the external circuit is utilized to generate an alternating current.

When coil and capacitor are connected in parallel

When a coil and a capacitor are connected in parallel, energy is stored at one instant in the charge coil (FIG. 3-6). Then the current is stored in the magnetic field around the coil and starts flowing the electrons to the other end of the capacitor plate to charge the capacitor, but in the opposite polarity. This cycle will be repeated as long as energy is supplied to the circuit. The above coil and capacitor operation is found in the hv section of the microwave oven.

3-6 The current is stored in the magnetic field when the coil and capacitor are connected in parallel (courtesy of Matsushita Electric Corporation of America).

HIGH-VOLTAGE CIRCUITS

Components operated from the hv power transformer may be considered part of the hv circuits (FIG. 3-7). The low ac power-line voltage is applied

NOTE: Door is closed.
Unit is not operated.

IMPORTANT SAFETY NOTE: THE SHADED AREA ON THIS SCHEMATIC DIAGRAM INCORPORATES SPECIAL FEATURES IMPORTANT FOR PROTECTION FROM MICROWAVE RADIATION, FIRE AND ELECTRICAL SHOCK HAZARDS. WHEN SERVICING, IT IS ESSENTIAL THAT ONLY MANUFACTURER'S SPECIFIED PARTS BE USED FOR THE CRITICAL COMPONENTS IN THE SHADED AREAS OF THE SCHEMATIC DIAGRAM.

3-7 Schematic diagram. Components operated from the hv power supply may be considered part of the hv circuit, from the secondary winding through the magnetron circuit (courtesy of Matsushita Electric Corporation of America).

to the primary winding of the hv transformer. High ac voltage from the secondary winding of the transformer forms a voltage doubling network with the hv capacitor and diode.

Two separate voltages are fed to the magnetron tube. A low 3 Vac is applied to the heater or filament terminals of the magnetron. Also, high negative voltage is fed to one side of the heater terminal and grounded anode terminal. Always discharge the hv capacitor before attempting to connect test equipment or placing your hands in the oven (FIG. 3-8).

3-8   Components found in the hv section of the microwave oven.

The hv circuits may be checked with low- and high-voltage test equipment. Although, some oven manufacturers do not recommend hv tests (for reasons of safety), proper hv test equipment and safety precautions provide adequate measurements in the hv circuits. The voltage applied to the hv transformer will indicate power-line voltage at the primary winding. Monitor the low voltage at the primary winding with an ac voltmeter or light bulb. An hv measurement at the diode and magnetron indicates that the transformer and voltage doubling circuits are functioning. A correct current measurement of the magnetron indicates the tube is operating.

When servicing intermittent ovens, monitoring the power transformer input and high voltage at the magnetron are critical measurements. Also, intermittent current readings may indicate a defective magnetron. Intermittent operation of the hv circuits may quickly identify a defective component with hv and current measurements.

THE CONTROLLER CIRCUITS

The electronic control circuits provide an easy method of operating the oven (FIG. 3-9). Simply tap out or push the time, temperature, and cook buttons for easy oven operation. The electronic controller circuit automatically controls the oven circuits and turns off the oven after the cooking time has expired. A digital display window shows the time while the

3-9 The electronic controller circuit provides an easy method to operate the oven. Simply tap out or push the various buttons (courtesy of Norelco).

digital programmer control circuit memorizes the function you have tapped out.

Most oven control circuits consist of a separate control board, digital clock display, and digital switch, while other ovens may have a microcomputer controlling the pulse oscillator, buzzer, display, humidity sensor, temperature power relay control, clock pulse, door check, variable power relay control, power relay control, membrane switch key board, and a dc power supply powering these circuits.

Check the electronic controller if there is no cooking operation after the proper sequence buttons are tapped out. Measure the ac transformer voltage going in and the control voltage going out of the electronic control board. The electronic controller operates a triac module or oven relay. A dirty switch assembly or poor cable connection may produce erratic oven operation. Improper digital numbers may be caused by either a defective digital display or electronic control board. Each control board may be exchanged or repaired when found defective.

DEFROST CIRCUITS

In the early oven circuits, a defrost motor rotates a defrost switch intermittently, causing the hv circuit to come off and on intermittently (FIG. 3-10). A defrost cycle with the electronic controlled circuits in some ovens provides an on/off time operation of a variable power switch. This variable power switch is located in the hv circuit (FIG. 3-11).

3-10 A defrost switch operates the hv circuits in an on-and-off manner. Turn the select switch to the defrost operation.

Check the setting of the select switch and defrost motor for possible defrost defects. Dirty contacts of the defrost motor assembly may produce intermittent defrost operations. Check the variable power switch for open switch connections when the defrost or magnetron circuits are not functioning. The variable switch controls may remain open resulting in no cooking action when in the defrost or regular cooking modes.

3-11 The variable power switch is located in the hv circuit. Intermittent voltage is applied to the magnetron by the control circuit.

Panasonic NN9807 cycle defrost When defrost power and defrosting time are selected and start pad is tapped:

1. The digital programmer circuit (DPC) divides the total defrosting time into eight equal periods, consisting of four defrosting periods, each followed by a standing period (FIG. 3-12).
2. During defrosting power periods, power relay B is energized for 9 seconds and de-energized for 13 seconds by DPC (FIG. 3-13).
3. During standing periods, power relay B is always open resulting in no microwave power.

3-12 Four defrosting and four standby periods equal a total of eight in the Panasonic NN9807 microwave oven (courtesy of Matsushita Electric Corporation of America).

3-13 Defrosting power periods that power relay B are energized for 9 seconds and de-energized for 13 seconds (courtesy of Matsushita Electric Corporation of America).

Defrost time selected is converted into seconds by the DPC, but display will show selected time in minutes and seconds as programmed. The total number of seconds is divided into eight time periods. The remainder (seconds not equally divisible by 8) are added to the last standing time period.

Panasonic MN-9507 variable power cooking control The coil of relay B is energized intermittently by the DPC when the oven is set at any power selection except for high power position. The DPC controls the on-off time of power relay B contacts in order to vary the output power of the microwave oven from warm to high power. One complete On-Off cycle of power relay B is 22 seconds. The relations between indications on the control panel and the output of the microwave oven is shown in FIG. 3-14.

INDICATION	OUTPUT POWER AGAINST HIGH POWER	ON, OFF TIME ON POWER RELAY B (RYB)
High	22/22 (700W)	22S ON / 22S ON
Med	17/22 (70%)	17S ON, 5S OFF
M. Low	13/22 (55%)	13S ON, 9S OFF
Low	8/22 (30%)	8S ON, 14S OFF
Warm	4/22 (15%)	4S ON, 18S OFF
Defrost	Approx 110W	Please refer to description of cyclic defrost.

3-14 Variable power cooking control power selection in a Panasonic MN-9507 microwave oven (courtesy of Matsushita Electric Corporation of America).

TEMPERATURE-CONTROL CIRCUITS

You may find a separate temperature-control circuit in manual control ovens. The temperature probe plugs into a probe jack inside the oven cavity. The thermistor probe controls the temperature control circuits and in turn controls the oven circuits for correct cooking modes. In some ovens, the select switch must be turned away from the temperature probe switch position for the oven to operate.

In ovens with a DPC control, the temperature probe is operated from the control circuits. The thermistor temperature probe jack connects directly to the electronic control board (FIG. 3-15). When the temperature probe is out of the probe jack, the short circuit contacts provide regular oven operation. The temperature probe circuits are controlled by the controller circuit board. Always remove the temperature probe when not in use.

3-15 The thermistor temperature probe jack connects directly to the electronic control board. Always remove the probe when it is not used. Do not lay the probe down on the metal oven floor when making oven tests.

Panasonic temperature control circuits The advantages of temperature cooking is being able to monitor the cooking temperature rise and eliminate the guess work of cooking. This circuit controls the temperature of food from 115 °F to 200 °F (or 46 °C to 93 °C).

1. When the temperature setting on the front panel is set to the desired temperature, the probe jack is connected to oven receptacle with the probe tip inserted into food, and the start button is pressed: approximately 18 volts generated in the tertiary winding of the hv transformer is converted to 24 Vdc by diode D1 and smoothed by capacitor C13 (FIG. 3-16).

2. The 24 Vdc is then decreased to 15 Vdc by R20 and is applied to the sensor circuit. The zener diode stabilizes the 15 Vdc. This voltage forward-brases the base-emitter junction of transistors Q7 and Q9 causing them to conduct (FIG. 3-17). Conduction of Q9 allows Q10 to open so that 15 Vdc forward-brases the base-emitter junction of Q11 and it conducts. The 24 volts is then applied to the power relay coil thereby energizing the power relay to close its contacts.

3-16 The dc voltage source fed to the temperature probe circuits (courtesy of Matsushita Electric Corporation of America).

3-17 The complete temperature probe circuit found in the Panasonic MQ8897BW oven (courtesy of Matsushita Electric Corporation of America).

These transistors can be considered as switching devices. The circuit consists of resistive dc voltage dividers that cause transistors to be either in a conduction or nonconduction state. There are no signal voltages in the circuit. The temperature control circuits are controlled by the microcomputer system.

3. When the food temperature goes up, the resistance of the thermistor goes down. Therefore, the voltage between base and emitter of Q8 goes up (FIG. 3-18). When the base voltage of Q8 exceeds its emitter voltage, Q8 will conduct. This happens when the temperature of food reaches a selected temperature.

4. When Q8 conducts, it causes Q9 to cut off and it causes Q10 to conduct.

3-18 How the transistors are biased and cut off during the temperature-probe cooking process (courtesy of Matsushita Electric Corporation of America).

5. When Q10 conducts, Q11 will cut off causing the power relay to de-energize and open the power relay contacts.

In this way, the temperature rise inside the food is controlled by the temperature circuit and the oven is automatically shut off when the food temperature reaches the desired temperature.

If a sensor probe component such as a thermistor or lead wire should go open, no bias is supplied to Q7 causing it to cut off. If Q7 cuts off, 15 Vdc forward-brases the base-emitter junction of Q10 through R21 and R32, allowing it to conduct and causes Q11 to cut off. If Q11 cuts off, it causes no power to be supplied to the power relay coil and the power relay shuts off the oven.

Construction of sensor probe The sensor probe is constructed with a thermistor, stainless steel pipe, lead wire, and jack assembly (FIG. 3-19). The probe is so constructed that no microwave leakage nor water leakage to the thermistor occurs so that the thermistor is not affected by such external conditions. It is recommended that the probe not be immersed into water to avoid possible water leakage into the thermistor portion.

CONVECTION CIRCUITS

A convection oven consists of a microwave and separate heater control circuit. The separate heating element provides cooking or browning procedures. Power may be applied to the heating element from a separate select switch or power relay (FIG. 3-20). The heating element may be controlled by the DPC unit. In some ovens, a thermostat switch may be located in series with the heating element to prevent overheating the oven cavity. Usually, a heater fan or convection motor circulates the hot air through the oven cavity.

You may find a large single heating element at the top of the oven for browning or cooking modes. In some early ovens, the heating element

3-19 Construction of Panasonic's temperature probe (courtesy of Matsushita Electric Corporation of America).

3-20 In a convection oven, power may be applied to the heating element from a separate selector switch or power relay. Also, the heating element may be controlled by the programmer circuit.

may be raised or lowered over the food to be cooked. While in other convection circuits you may see a round heating element with a fan motor in the center of the element, these circular heating elements are generally located at the center and top area of the oven cavity.

Panasonic convection cooking control The digital circuit controls the on-off time of the heater in order to control oven cavity temperature.

1. After the start pad is tapped with the desired convection program set, an 18 Vdc signal comes out of the DPC and is applied to coil of power relay C.

2. When the contacts of power B close, power-source voltage (120 Vac) is applied to the heater and the heater turns on.

3. When the oven temperature reaches the set temperature, the DPC senses the temperature through oven temperature sensor and stops supplying an 18 Vdc signal to the coil of power relay C and the heater turns off.

4. After the heater turns off, the oven temperature will continue increasing awhile and then decrease (FIG. 3-21). When the oven

Oven temp	Start ▼ Selected oven temp Oven temp
Power relay C	On | Off | On | Off
Heater	On | Off | On | Off

3-21 In a convection oven, power may be applied to the heating element from a separate switch or power relay (courtesy of Matsushita Electric Corporation of America).

temperature drops below the set temperature, the DPC senses the signal and starts supplying an 18 Vdc signal to the coil of power relay again.

In the Panasonic NN-8907 oven, an upper and lower heating element is located. Here power relay D and C closes the contacts for the heating elements (FIG. 3-22). The DPC controls four different power relays. Only the top element is used in the broil mode. The upper and lower heating elements alternately go on and off in accordance to the cooking temperature. Some of these ovens have a thermal cutout in series with the heating element to remove the element from the circuit.

THE COOKING CYCLE

Before attempting to service the microwave oven, make sure you know how to operate the oven controls. The manual operated oven is very easy to operate, while some electronically controlled ovens are more difficult. Always have the operation manual of the oven handy when testing or using the oven. Not only will the oven operation manual start you off in the right direction, but it may help you locate the defective component. A typical normal cooking sequence of a manual operated oven may begin as follows:

1. Place the food to be cooked in the oven and close the door. When the door is closed all interlock switches are closed except the monitor or safety switch. The monitor or safety switch opens with the door closed (FIG. 3-23). Follow the dark bold lines in the schematic diagram. Here the oven is in the idle or off condition until the timer is set.

2. Next, turn the timer dial for the desired cooking time. With the door closed and the timer set, the fan blower motor will start to

IMPORTANT SAFETY NOTE: THE SHADED AREA ON THIS SCHEMATIC DIAGRAM INCORPORATES SPECIAL FEATURES IMPORTANT FOR PROTECTION FROM MICROWAVE RADIATION, FIRE AND ELECTRICAL SHOCK HAZARDS. WHEN SERVICING, IT IS ESSENTIAL THAT ONLY MANUFACTURER'S SPECIFIED PARTS BE USED FOR THE CRITICAL COMPONENTS IN THE SHADED AREAS OF THE SCHEMATIC DIAGRAM.

3-22 Power relay D and C place the upper and lower heating elements in the circuit for broiling and baking modes (courtesy of Matsushita Electric Corporation of America).

3-23 The monitor or safety switch opens with the door closed. When the door is closed, all interlock switch contacts are made.

run. The air starts to circulate in the oven and cool the magnetron. The oven light will come on (FIG. 3-24). Notice power is applied to the cook switch.

3-24 With the door closed and the timer set, the fan motor will start to operate. The oven light will come on.

3. Push the cook button. The cook indicator light will glow indicating the oven is in the cooking cycle. When the cook switch is depressed, the timer motor begins to operate (FIG. 3-25). Power-line voltage is found across the cook light and the primary winding of the hv transformer. High ac voltage from the secondary of the power transformer feeds into a voltage-doubling diode and capacitor network with a high negative voltage applied to the heater terminals of the magnetron. Low ac voltage is applied to the heater or filament terminals. The magnetron begins to oscillate providing rf energy to cook the food in the oven cavity.

3-25 When the cook switch is depressed, the timer motor begins to operate. Now, the power-line voltage is found across the cook light and primary winding of the hv transformer.

4. After the end of the cooking cycle has expired, the timer control opens up removing the power-line voltage from the rest of the circuit. The timer returns to 0 on the dial. A timer bell sounds, indicating the end of the cooking cycle. After the cook cycle has ended the food is removed from the oven. When the door opens the monitor or safety switch closes contacts and the other interlock switches open up. Remember the safety switch contacts are closed with the door open (FIG. 3-26). This prevents the oven from operating and will blow the 15 amp fuse if any of the other interlock switches hang up. Knowing the cooking sequence with each component in operation is helpful in locating the defective component. Follow the schematic as each component begins to operate. When the next cooking sequence ceases to function, suspect the corresponding component.

3-26 Remember—the safety switch contacts are closed with the oven door. Always check the condition of the switch after the 15 amp fuse blows.

A typical microwave oven cooking sequence with a digital electronic controller is given as follows.

1. When not cooking with the temperature probe, make sure the probe is unplugged and out of the oven.

2. Open oven door and place food or water test in the oven cavity. With the oven door closed, the low voltage transformer supplies ac voltage to the control circuit as the power cord is plugged in. When the door is closed, the contacts of the safety switch are open. The latch or interlock switch contacts are made to furnish power-line voltage to the low-voltage circuit. The oven light turns on. In some ovens, the oven light is on when the door is opened or closed (FIG. 3-27).

3. Tap the cooking time and power controls of the digital control board. The power indicator light turns on to indicate power has been set. The time appears in the digital display window. Also the control circuits memorizes the functions you have set.

4. Tap the start or cook control button. The coil of the power or oven relay is energized by the control board. In other ovens, the

3-27 In some ovens, the oven light is on when the door is opened or closed, while in other ovens, the cavity light is only on when the timer or programmer is in operation.

triac assembly is energized by the control board. Power-line voltage (120 Vac) is applied to the primary winding of the hv transformer through the relay contacts or triac assembly (FIG. 3-28).

The fan motor begins to operate blowing air against the magnetron and exhausting vapor through the vent areas. In some ovens, the same fan motor may rotate the stirrer blade with a pulley-gear belt arrangement. In other ovens, you may find a separate stirrer motor. In some Sharp ovens the turntable starts rotating.

A power indicator light may start blinking to indicate the oven is functioning with the oven lights on. The cooking time starts to count

3-28 Power-line voltage is applied to the hv primary winding through the relay contacts or triac assembly. They are energized with the start or cook button of the digital control board.

down in the digital display window. The power-line voltage applied to the primary winding of the hv transformer, filament or heater voltage is applied to the magnetron with high ac voltage fed to the voltage-doubler network. The hv capacitor and diode combine to form the dc voltage doubling circuit. A negative high voltage is applied to the heater circuit of the magnetron. The magnetron begins to oscillate with the rf energy being emitted through the waveguide assembly to the food in the oven cavity.

The stirrer motor or stirrer-fan motor is found in several ovens, spreading rf energy to prevent hot spots in the oven and to prevent uneven cooking. The lower motor has a pulley with a chain type belt that rotates the blade, while the stirrer motor is a direct drive type. The rotary blade may be called a rotary antenna in the Amana and radiating antenna in the Hardwick ovens (FIG. 3-29).

3-29 The rotary antenna blade may be located in the top of the oven. Sometimes a drop or two of oil on the fan or rotary pulley will quiet down the squeaking noise.

In some of the ovens, including Sharp and Samsung, a turntable rotates the food for even cooking. The motor for the turntable has a reduction gear-box in the bottom of the oven. You may find a variable power switch or relay in the hv circuits. It is energized intermittently, often with a digital control circuit.

With this knowledge of operating the oven, how the cooking sequence functions, and closely following the circuit diagram, you may quickly locate a defective component. First, isolate the service problem with the circuit diagram. Next, see how far into the suspected circuit various components are actually working. Take voltage and resistance measurements in locating the defective component. Then, replace the defective part.

REMOVING THE WRAP OR COVER

Before any repairs can be made, the back cover or wrap must be removed. First, disconnect the oven from the power-line plug. Do not leave the oven plugged in when removing the outer case. You may find several screws at the back along the top and side edges. Check for additional Phillips metal screws along the bottom edge of both sides of the outer case or cover (FIG. 3-30).

3-30 Remove all screws around the outside rim at the back cover so the entire panel will lift off the oven unit (courtesy of Norelco Corp.).

In older and commercial type ovens, you may find several back panels that should be removed to get at the various working components. The wrap-around metal cover may be welded to the metal bottom plate. Servicing these type ovens may take longer since the components are difficult to reach.

When all cover screws have been removed, slide the entire case back about one inch. You may have to lift up the back side of the case to free it from retaining clips on the cavity face plate. Set the outer cover out of the

3-31   After removing the cover, discharge the hv capacitor before making any test or part replacement (courtesy of Norelco Corp.).

way so it will not get scratched up. Before touching or checking any component, discharge the hv capacitor (FIG. 3-31).

After all oven repairs have been made with cook and leakage tests, replace the outer case. Make sure the power cord is pulled. Double check the top and side front edges. Be sure the edges are tucked inside before replacing the metal screws. If not, the case may vibrate and light may show through the seam areas when operating. Clean off the outside case and control areas with window cleaner or a mild detergent and water.

Chapter 4

Basic troubleshooting procedures

*B*efore attempting to service any microwave oven, you should know the trouble symptoms, have the correct test instruments, know certain test procedures, secure the oven schematic, and use a lot of common sense. The more you know about the trouble symptoms, the easier it is to locate the defective component. The owner should be asked several questions about how the microwave oven operates:

- What were you doing when the oven quit?
- Is the oven dead or intermittent?
- At what point in the cooking cycle did the oven stop operating?
- Is there anything more that you can tell us about the oven?
- How old is the microwave oven? (Date purchased.)
- Is the oven in warranty? (Warranty registration card.)
- Where was the oven purchased? (Name and location.)

In case the oven is in warranty, make sure you see the warranty registration card or bill of sale. Most ovens have a one or two year labor warranty—two years on parts, and from five to seven years warranty of the magnetron tube. Besides the trouble symptoms, you should list the model and serial number of the oven on the service repair tag. List the date and place where the oven was purchased. At this point, you should tell the customer if he is to pay labor, parts and labor, or if the entire repair is covered by the warranty. This data must be listed on the warranty repair order to collect payment on warranty service.

OWNER'S RESPONSIBILITY

Many of the troubles caused in the microwave oven are caused by the operator of the oven. Make sure the oven is plugged into the outlet receptacle. If the oven does not heat properly and is connected to an improper extension cord, check for proper ac voltage at the oven plug (117−120 Vac). No voltage, check the house fuse. A separate outlet should be installed for the microwave oven with at least number 12 wire and properly grounded (FIG. 4-1).

4-1 Check the ac wall outlet with a meter, floor lamp, or small kitchen appliance to see if the outlet is alive. Check the house fuse with no voltage at the outlet.

Know how the oven operates. Read the oven instructions starting and cooking tests over several times. Improper oven operating results in another service call. Have the service technician or salesman go over the instruction manual in detail. Do not be afraid of any buttons on the oven.

Do not be afraid of radiation from the oven. If the front glass accidentally gets cracked or the door will not close properly, pull the plug of oven and call the electronic service technician. Remember, the oven technician should check the oven radiation before the oven is returned to the owner. Have the technician check for radiation of oven in the home if you are still in doubt.

WHERE TO START

Take the list of trouble symptoms with the service literature, isolate the trouble, locate the defective component, and replace the new part (FIG.

4-2 Schematic diagram of a Norelco MCS8100 microwave oven. Start troubleshooting with a schematic, trouble symptoms, and isolation procedures (courtesy of Norelco Corp.).

4-2). Turn to the service diagnosis and trouble symptom charts in the service literature. Most oven procedures start at the beginning and each stage is checked off as you go through the various test procedures. When the possible malfunction occurs, test the various components listed.

The oven symptoms may be listed as low voltage, control, and high voltage (hv) circuits. Any possible symptom or trouble that may occur with the power-line voltage (115 or 120 Vac) may be listed as low voltage circuits. The control circuits are either manual or electronic board control circuits. The manual circuits are controlled by the oven switch and timer while the electronic control circuits are controlled by the electronic push button type control board. All hv circuits exist from the power transformer through the magnetron tube circuit.

ISOLATION

After knowing the oven sequence of operation and possible symptoms you can quickly eliminate the defective components. Simply check off each working component according to the symptom. When you reach the dead area, use the schematic to point out the possible defects. The following are several actual microwave oven problems, broken down with a partial schematic to demonstrate each trouble.

DEAD CONDITION

The oven door is closed, timer set for three minutes, and the oven cook switch is pushed—nothing happens. The oven appears dead, even the oven lamp does not come on and there is no fan rotation. What is the trouble? Let's take a breakdown of a simple microwave oven circuit (FIG. 4-3).

Of course, you remove the fuse the first thing and check continuity with the ohmmeter. Replace the fuse if it is open. Now, see if the oven operates. The 15 amp fuse opens up when there is an overloaded condition in the oven. Maybe one of the interlock switches did not open up when the door was opened. Perhaps the magnetron or hv diode became leaky or shorted out. It's possible the power line may be overloaded causing the fuse to open. Sometimes the fuse is blown when the operator rapidly changes the timer and push button control board without shutting the oven door. Often, the open fuse symptom will show up the possible defective component in a four-hour cooking test.

The fuse tested good but you replaced it anyway, so what could be the trouble? Glancing at the schematic, the timer switch must be defective if the blower motor doesn't start to operate. If the fan motor is operating and there is no oven light, the thermal cutout on the magnetron must be open. Pull the ac power cord. Take an ohmmeter reading across the thermal cutout contacts. When normal, you should have a shorted condition with no resistance measurement.

The oven light should come on when the timer switch contacts are closed or the timer is set for a few minutes on the dial. Check the timer

4-3 The oven lamp should come on if the fuse, timer, thermal cutout, and light bulb are normal. Notice the dark lines of the circuit and the components within the circuit to let the oven lamp become an indicator.

4-4 Check the interlock switch assembly with the ohmmeter when the oven light does not come on. Open and close the door to see if the meter responds.

switch contacts and thermal cutout assembly with the low ohmmeter range. Both components should show a shorted or closed condition. If the oven light now is on but the oven still does not function, suspect a defective interlock switch, cook switch, and cook relay (FIG. 4-4).

OVEN LIGHT ON BUT NO COOKING

When the cook switch is pushed, the oven relay should energize and apply ac to the primary winding of the power transformer (FIG. 4-5). Follow the double-lined wires in the schematic. We know that ac voltage is at point (1) to the blower motor and the oven light. The oven relay must be energized to have voltage at point (2). Listen very carefully, you can hear the cook relay working when the cook switch is pushed. A thud or jamming noise indicates the relay is operating (FIG. 4-6).

Since the cook relay is not energizing, suspect the cook switch contacts, interlock switch 2, or an open cook relay solenoid. Always check the interlock switches after the fuse for possible trouble. They cause more problems than any other component in the oven. The interlock switches are closed and opened each time the oven door is opened. So the likely component is interlock switch 2. Locate switch 2 in the service literature (FIG. 4-7).

The contacts of interlock switch 2 are normally open when the door is opened. Once the door is closed, the contacts of interlock switch 2 are closed. With the door closed, clip the ohmmeter across the two switch contacts. You should have a shorted measurement (FIG. 4-8). If not, the switch is open. Now open the door and notice if the meter hand will show a shorted reading and open when the door is opened.

4-5 When the cook switch is pushed, the oven relay should energize, applying the ac to the primary winding of the power transformer. Follow the dark line wire hookup in the drawing.

4-6 You can hear the oven cook relay working when the cook switch is pushed. A thud or jamming noise indicates the relay is operating.

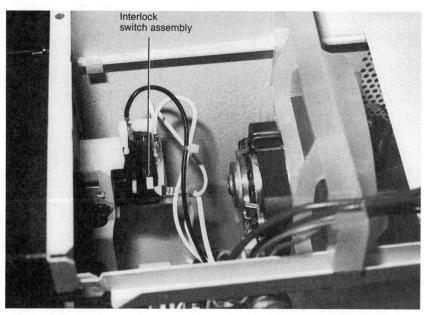

4-7 Remove the suspected interlock switch and check the circuit with the ohmmeter. The switch may be open, intermittent, and/or erratic. Replace it if you're in doubt.

4-8   Locate the correct interlock switch on the schematic and in the oven. Interlock switches may be checked by removing the leads and clipping the ohmmeter across the terminals. Set the meter to R×1. A normally closed (NC) interlock will show a shorted measurement.

In case the switch 2 is normal, check the cook switch and timer switch contacts for closed contacts. When the cook switch is pushed, the timer switch contacts will close with the timer set at a few minutes on the timer dial. Sometimes this timer switch contact may be erratic in operation, causing the oven to work one minute and not the next. Intermittent or erratic oven conditions are a little more difficult to locate.

In case interlock switch 2 is normal, check the switch contacts with the ohmmeter. Make sure the ac power cord is disconnected. Clip the ohmmeter leads across the oven switch terminals. Push in on the oven switch. Most oven switch assemblies have momentary contacts. Contact is made only when the switch is pushed. You should measure a dead short across the normally open switch. Let up on the oven switch and the meter reading should be open.

The only component left is the oven relay. There are two ways to check the oven relay. Measure the ac voltage (120 V) across the relay solenoid terminals or take a resistance reading. When ac voltage is across the solenoid winding with no movement of the relay, suspect an open winding. Double check with the ohmmeter.

With the dead condition test, we are using the oven lamp and fan motor as indicators to breakdown the operating circuits. It's possible to have a defective oven lamp, but in some ovens there are at least two different bulbs. Since the oven light is now on, we know part of the circuit is operating to point (1). Also, with the fan blower rotating, we know that voltage is normal this far in the circuit. It must be said here that all fans and oven lights may not be wired the same in different microwave ovens. By using different components as indicators, we may quickly find the defective component.

In ovens with tab button or electronic control circuits, the electronic control unit must operate before the oven light comes on. The electronic control circuits may operate a power relay or a switching device (triac). The electronic control circuits and triac must function before the oven light comes on in a Panasonic oven (FIG. 4-9).

OVEN LIGHT ON, COOK RELAY ENERGIZED, NO COOKING

We know the relay is energizing because we can hear it close and also the oven light is on, indicating voltage at point 2. To determine if voltage is getting to the power transformer, measure the ac voltage across the primary terminals. These terminals are usually enclosed in plastic covering and should either be pulled back or use needle-type test probes which go into the wire insulation. It's best to clip the ac meter across the primary winding of the power transformer.

Since ac voltage is found at point 2 of the cook relay, suspect interlock switch 1 for being open or erratic in operation (FIG. 4-10). Both switch 1 and the monitor or safety switch should be replaced when one is found to be defective. Locate the interlock switch 1 close to the front door latch. Normally, the switch is closed when the oven door is shut and opens up with the door open. Clip the ohmmeter leads across the termi-

4-9 Both the control circuit and the triac must function before the oven lamp comes on in this Panasonic oven (courtesy of Matsushita Electric Corporation of America).

4-10 Check for interlock switch I open connection when the cook relay energized and no ac is applied to the primary winding of the power transformer. Switch I is in the other leg of the line opposite from the cook relay contacts.

nals. Open and close the door and notice the ohmmeter readings. Interlock switch 1 and 2 and the 15 amp fuse produce more dead and intermittent operation than any other oven components.

NO TRANSFORMER VOLTAGE

Clip the ac voltmeter or test lamp across the primary winding of the power transformer. If the lamp is on and 120 Vac is measured at the transformer, suspect a defective component in the hv circuits. Suspect poor power relay contacts or a defective triac and control circuit (FIG. 4-11). Does the fan blower come on? The fan blower is another indicator showing that ac voltage is applied this far in the circuit.

4-11   Suspect poor oven or power relay contacts, a defective triac, or a control unit with no voltage at primary winding.

TRANSFORMER VOLTAGE BUT NO COOKING

In some ovens, you may find a defrost timer motor and indicating lamp that is wired across the primary winding of the power transformer (FIG. 4-12). When the defrost mechanism is operating you may assume ac voltage is applied to the large power transformer. In other ovens, the fan motor may be tied across the ac primary winding. Look for working components through the schematics to help locate the various troubles.

After a normal voltage measurement is found on the primary winding, with no operation of the magnetron, suspect a defective component in the hv section. Before taking any voltage or resistance measurements in the hv circuit, discharge the hv capacitor each time the oven is fired up. This hv capacitor may hold a charge unless a high-megohm resistor is found across the capacitor to leak off the voltage.

The most likely defective component in the hv circuits are the hv diode and magnetron tube (FIG. 4-13). Correct hv measurement at the magnetron tube indicates the hv circuits are normal and the magnetron tube must be defective. No or low voltage at the magnetron tube may

4-12   If both the defrost timer motor and the lamp are functioning, this indicates voltage should be at the power transformer. Check the defrost cook switch for possible bad contacts. Do not overlook open primary winding or poor ac connections.

4-13   Check for ac voltage across the secondary winding to chassis ground. Remember this voltage is high (1.8 to 3kV). Clip the voltmeter or Magnameter onto the circuit with the plug removed from the outlet.

result from a leaky magnetron tube or any component in the hv circuit. Proper high voltage with no current measurement at the magnetron tube indicates a defective magnetron.

Measure the ac high voltage across the secondary with the Magnameter or a hv probe. Remember this voltage may be from 2kV to 3kV across terminal numbers 3 and 4. A very low ac voltage measurement may be caused by a leaky capacitor, diode, or magnetron tube. The filament voltage may vary between 3.15 and 3.2 Vac (FIG. 4-14). Be careful as these filament leads are at a high dc voltage potential to chassis ground. Information about servicing the hv and low-voltage circuits are given in chapters 6 and 7.

TYPICAL TROUBLESHOOTING CHARTS

Here are two different microwave oven troubleshooting charts. Chart I is for a typical manual-type timer oven with possible defects (TABLE 4-1).

4-14 The filament voltage should not be measured, although it is only 3.3 Vac. Remember, high dc voltage is on these terminals to chassis ground. You may find a separate power transformer for the filament voltage in some ovens.

Chart II is a typical troubleshooting chart for an electronic control board (TABLE 4-2). Most troubleshooting charts provided by the manufacturer are about the same, but each one may have a different method of servicing. If a service manual is handy for the particular oven you are servicing, use it. You may want to take a few notes of a given trouble and list them right in the schematic for future reference.

TYPICAL OVEN SYMPTOMS
AND POSSIBLY DEFECTIVE COMPONENTS

All of these symptoms are actual troubles that have been found while servicing microwave ovens. Each symptom is given with the actual breakdown of the component. This list is given to help to quickly locate a definite symptom with the possible defective component. These problems are not listed in any special order or as they occurred.

Dead—Nothing works

1. Open 15 amp fuse
2. Bad latch switch
3. Replace open interlock switch
4. Broken door latch spring, will not activate interlock switch

Table 4-1   A typical manually-operated oven with possible defects.

Problem	Possible Defective Component	Test and Replace
House fuse blows when power cord inserted	Shorted power cord & plug	Replace ac power cord
Fuse blows when open the door	Check interlock monitor and safety switch	Replace both safety and interlock switch
Fuse blows when oven turned on	Check interlock switch Check line varistor Check safety switch	Replace interlock switch See if shorted See if line varistor arcing - Replace Hang up safety switch Replace
Fuse blows when cook switch is pushed	Check hv capacitor Check hv diode Check magnetron Check power transformer	Test and replace the defective component
Oven dead with timer and cook button on	Check for blown fuse Check for power in home Check defective timer Check component in hv circuitry Bad latch or interlock switch Defective thermal-cut out	Replace with 15 amp chemical fuse Measure with hv voltmeter Replace timer Test all hv components Check continuity and replace Check continuity
Blower motor won't rotate	Bad pligen connection Poor soldered cable	Check and reset ac plug Check voltage at motor terminals
Blower motor won't rotate	Defective motor	Check motors for dry or gummed up bearings
Oven light does not come on	Burned out bulb Defective timer contact Defective thermo-cut out Poor wiring to socket	Replace bulb Check timer switch with ohmmeter Measure across terminal with ohmmeter, should have short reading Visually inspect wiring
Oven goes into cook cycle - timer does not operate	Defective timer motor Open wiring to timer	Replace timer assembly Check for ac voltage (120V) at timer motor terminals
Timer erratic - sometimes stops	Defective timer assembly	Replace entire timer assembly
Oven cooks - no cook light	Defective cook light Bad wiring	Replace cook light - some of these are new type bulbs Check for 120 Vac across light bulb terminal assembly
Oven light on - No heat or cooking	Defective power transformer Defective hv diode Defective hv capacitor Defective magnetron	Check voltage at primary 120 Vac Check if warm or leaky Open or shorted - replace if in doubt Measure hv and current of magnetron tube

Table 4-2   A typical troubleshooting chart for an oven with an electronic control board.

Problem	Possible Defective Component	Test and Replace
Oven light out when door opened	Defective interlock switch Open or poor wiring Defective light	Check with ohmmeter with door open and closed Visually inspect wiring Replace bulb
Oven light does not light at all	Burned out bulb Poor wiring to light socket	Replace - check maybe 2 bulbs in oven Check and repair wiring
Oven ready with water test - components stop operating when door closed	Defective oven relay Defective triac Display stops and doesn't count down	Check solenoid with ohmmeter Check contacts Check as in Chapter 6 Check door sensing switch Check wiring Check low voltage transfer Replace defective electronic circuit control board assembly
Cook indicator light on - No fan motor rotation	Check door interlock switch Check door alignment Defective Auxillary relay Defective blower motor	Check continuity with ohmmeter while opening and closing door Re-align door Check relay
Cook indicator light on - No fan motor rotation	Broken wires to motor	Measure voltage at fan motor Lubricate bearings Check wires for poor connection or plug end Measure ac voltage at motor terminals
Turntable motor doesn't rotate	Defective turntable motor Check wiring	Measure voltage at motor terminal See if motor coupling binding Visually inspect wiring and voltage Replace motor if defective
Stirrer motor does not rotate	Check defective motor Belt off fan blade	Check motor continuity and voltage Visually inspect belt and replace

Table 4-2 Continued.

Problem	Possible Defective Component	Test and Replace
First steps times out - tone sounds oven shifts to memory time - color begins to flash - tone sounds three times	Check for improper oven grounding Low line voltage (below 110V) Defective electronic contact board assembly	Check continuity of grounds with ohmmeter and voltmeter Measure ac line voltage Notify electric company if low Replace electronic contact board assembly
Oven Operates but low heat or heats slowly	Check line voltage (below 110V) Defective power transformer Defective hv capacitor Defective thermal cutout (intermittent) Defective magnetron Defective cook relay Defective triac Defective control unit	Measure ac line voltage should be 115 to 120 Vac Check primary & secondary voltage Take winding resistance test Replace Place ac meter across thermal contacts when in question Take high voltage and current measurements Check continuity of solenoid Check for leaky conditions - replace Replace
Oven erratic cooking	Defective touch panel Defective triac module Defective electronic control board assembly	Replace Replace Replace
Oven goes into cook cycle - uneven heating or cooking	Stirrer blade does not rotate Burned waveguide cover Turntable does not operate Defective magnetron	Check stirrer motor or belt assembly Replace cover Check ac voltage on motor terminals Replace broken coupling bushing Replace after voltage and current tests
Using temperature probe oven does not function	Switch not turned to probe Temperature probe open or shorted Defective probe jack Defective electronic control board assembly	Check switch control setting Check temperature probe with ohmmeter Check jack contacts with ohmmeter Replace control board assembly

Problem	Possible Defective Component	Test and Replace
Push temperature pad center 140 - close door oven begins to operate oven beeps when pad is touched	Defective probe Defective probe jack	Check with ohmmeter Check with ohmmeter

4-15 You may find a defective control board causing no heat or oven to come on. Here is a backside view of an electronic control board of a Sharp R5A50 model.

5. Realign and adjust door

6. Replace switch and monitor assembly

7. Only oven light operation—control set to probe—no probe inserted

8. Loose plug or power board

9. Replace defective control board (FIG. 4-15).

10. Defective oven relay

11. Bad wire connections—broken

12. Open thermal cutout

13. Upper door switch

14. Open low-voltage transformer

15. Poor transformer primary connection
16. Leaky hv diode
17. Defective triac
18. Shorted hv capacitor
19. Defective magnetron tube

Intermittent operation

1. Bad ac plug in house
2. Defective extension cord
3. Intermittent latch or interlock switches
4. Poorly crimped connection on power transformer leads
5. Intermittent control board (replace entire board assembly)
6. Defective magnetron tube
7. Improper setting of door interlock switches
8. Door realignment
9. Defective thermal cutout

No heat—No cooking

1. Leaky magnetron
2. Leaky diode
3. Defective thermal cutout
4. Defective triac
5. Replace upper switch strike assembly
6. Shorted hv capacitor

Slow cooking

1. Defective magnetron tube
2. Leaky diode (just runs warm)
3. Defective triac assembly (FIG. 4-16)
4. Change electronic control board
5. Shorted power transformer
6. Overheated thermal cutout
7. Improper or low ac line voltage

Erratic cooking

1. Defective magnetron
2. Poor transformer crimped connection
3. Poor connection to control board

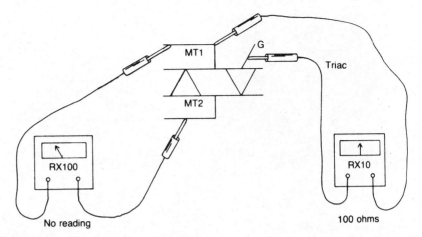

4-16 A leaky or shorted triac assembly may cause slow cooking in the microwave oven. Check the triac with the ohmmeter, diode tester, or triac checker.

4. Defective control board
5. Hv diode
6. Overheated thermal cutout on magnetron
7. Erratic interlock or latch switch
8. Dirty oven or on/off switch
9. Broken door spring to engage interlock switch

Lights up—No fan—No cooking

1. Defective interlock and latch switches
2. Defective control board assembly
3. Door alignment

Everything operates—No cooking

1. Defective magnetron tube
2. High voltage present—no current—tube defective
3. No hv—defective capacitor or leaky diode
4. Shorted or open power transformer

Can't shut oven off

1. Leaky triac (FIG. 4-17)
2. Defective oven relay

No cooking—Only a hum noise

1. Replace leaky triac assembly

4-17 A leaky triac assembly may cause the oven to keep operating. In some ovens, a defective diode may not allow the oven to turn on.

Loud pop and oven quit

1. Leaky or shorted capacitor (FIG. 4-18)
2. Leaky or shorted hv diode

4-18 A leaky or shorted hv capacitor may cause the oven to pop and quit. Here the resistance between the two capacitor terminals is only 1.4 ohms. No resistance should be measured across a normal hv capacitor.

Keeps blowing line fuse

1. Hung up interlock switch
2. Loose door and latch assembly
3. Requires door alignment (too much door play)

4. Leaky hv diode
5. Shorted magnetron
6. Arcing ac line varistor

Blows fuse when open door

1. Replace interlock switch and monitor or safety switch
2. Realignment of door

Open door—Went dead

1. Replace defective monitor and interlock switch
2. Replace control board assembly
3. Replace oven relay
4. Replace defective triac

Oven stops operating when door is closed

1. Blown fuse
2. Defective interlock switch (FIG. 4-19)
3. Defective oven relay
4. Defective triac
5. Open low-voltage transformer

4-19 Check for a defective interlock switch when the oven is closed. When the safety or monitor interlock switch blows the fuse, replace the entire switch assembly in the Sharp microwave ovens.

6. Defective touch panel control board
7. Defective control board
8. Realignment of door

Oven runs too hot

1. Defective magnetron
2. Too high ac line voltage (above 130 Vac)
3. Burns food—defective magnetron

Something burning in oven

1. Never use regular paper bags for cooking
2. Grease behind shelf guide and metal screws
3. Grease behind waveguide cover
4. Replace waveguide cover

Hot spots

1. Improper cooking
2. Defective magnetron tube

Arcing in oven

1. Exposed or loose oven bolt
2. Loose front door screw
3. Grease behind shelf metal screws
4. Loose nut or screw on browning element inside oven
5. Waveguide cover (replace)
6. Sparking from defective magnetron tube

Very noisy

1. Fan assembly loose (FIG. 4-20)
2. Noisy pulley or stirrer assembly
3. Constant buzzing of the power transformer (replace)
4. Noisy fan (needs lubrication)
5. Loose fan bracket
6. Noisy timer motor assembly (lubricate or replace)
7. Noisy turntable motor (lubricate or replace)

No fan operation

1. Fan blade jammed
2. Poor socket from fan to control assembly
3. Bad fan motor lead (poorly soldered connection)
4. Won't turn—frozen fan, needs lubrication

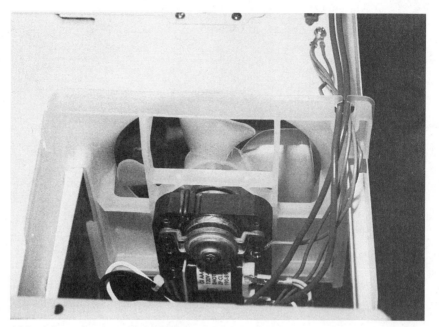

4-20 A loose fan assembly may let the fan blade touch the sides and cause a vibrating noise. Realign the fan and tighten the mounting bolts.

   5. Intermittent fan operation—bad cable plug
   6. Slow coming on—needs lubrication

Door doesn't close properly

   1. Door alignment and adjustment (FIG. 4-21)
   2. Replace damaged door

4-21 Check the door for poor alignment and adjustment. Remove four screws holding the door in a Panasonic MZ8877BW oven (courtesy of Matsushita Electric Corporation of America).

3. Door sticks—realign door guide assembly (FIG. 4-22)

4. Replace loose screw in switch housing assembly

Loose door latch

1. Broken spring in door

2. Loose screw

3. Loose pop-rivet

No oven light

1. Defective bulb

2. Defective socket

3. Defective oven switch

4. Wire off/on cable plug

5. Oven light doesn't go off—replace latch or oven switch

Timer runs slow or erratic

1. Replace timer assembly (FIG. 4-23)

Timer cord will not operate oven

1. Replace cord assembly

2. Check cord inlet assembly

Turntable doesn't operate

1. Replace broken turntable plastic coupling

2. Replace turntable motor

3. Check wiring on motor (measure ac voltage)

4. Check turntable bearings

No defrost cycle

1. Replace defective defrost switch

2. Replace defrost timer motor assembly

Browner doesn't function

1. Defective switch

2. Open heating element

3. Bad flexible wire connection at end of heating element

Oven shock

1. Check for three-prong ac grounding plug (FIG. 4-24)

2. Check grounding of oven

4-22 When the door doesn't close properly, check the door for proper adjustment. Readjust the hinge mounting nuts as shown in the Norelco MC8100 oven (courtesy of Norelco Corp.).

HINGE

NYLON SPACER

HINGE MOUNTING NUTS

HINGE

NYLON SPACER

4-23 When the timer becomes slow or erratic, replace the entire timer assembly. You may find a defective timing unit will stop in the cooking cycle. Here is manual timer from a Goldstar microwave oven.

4-24 Check the grounding of the oven for possible oven shock. Measure ac voltage between the ground plug on the outlet and the oven cabinet.

3. Static elements between kitchen carpet and oven
4. Discharge static electricity in oven metal before touching control panel (may damage IC components in touch control board)

PROPER POLARIZATION AND PROPER GROUNDING

For the safety of oven operators and service technicians, the microwave oven with a three-prong plug must be grounded and polarized at the wall outlet. The round hole should be grounded at the receptacle through 3 wire cable back to the switch box. Polarization means the larger slot of the receptacle must be neutral and the small slot the hot wire (FIG. 4-25). Check the color of the wire screws. The small slot (hot) should have copper colored screws and the large slot (neutral) should have white colored screws.

4-25   Make sure the three-prong plug and the outlet are connected properly.

Do not cut off the ground round terminal of the ac oven plug if the outlet will not accept it. Use a grounding adaptor temporarily until the electrician can run a new three-wire cable to the outlet. All microwave ovens should be operated from a separate outlet to prevent overloading and proper oven operation.

CONCLUSION

The results of basic troubleshooting procedures are to take the trouble symptom, diagnosis, and then isolate the suspected component, or locate the suspected component from the manufacturer's diagnosis service procedures or typical trouble procedures. Isolate or breakdown the circuits where the trouble may occur. Test the suspected component with those given or found in the service literature. Replace the defective part or component with the original manufacturer's replacement parts.

While trying to isolate and service the various microwave oven circuits you may find more than one trouble within the oven. For instance, you may have a blown fuse, defective interlock switch, and electronic control board. When lightning or power line outages occur, you may locate several different damaged components. Excessive lightning charges may cause the fuse to keep blowing because of an arcing power-line varistor. Lightning charges can also damage the electronic control board, power switches, and small power transformer.

Triac checker

This little tester will quickly test the condition of the triac assembly found in the microwave oven. The checker consists of only 10 small parts that can be picked up at the local Radio Shack store. No on and off switch is found as the tester can be plugged into the ac outlet when needed. Build the checker into a small plastic box so it can fit into the tool caddy.

The circuit

T1 is a 12.6 Vac step-down power transformer that provides ac voltage to a halfwave rectifier D1 (FIG. 4-26). After the voltage is rectified to dc, C1 smooths out the ripples for the triac circuits. R1 provides dc voltage to the MT2 terminal. R2 limits the dc voltage applied to the red LED (light emitting diode) indicator. SW1 is a momentary open push button switch, while R3 (50k) resistors varies the voltage applied to the gate terminal. MT1 connects to the negative side of the power supply.

4-26 Circuit diagram of the triac checker. Notice there are only a total of 10 parts.

Testing

Remove the leads from the triac to be tested. Clip the black alligator clip to the MT1, red clip to MT2, and green lead to the gate terminal of the triac. The normal triac will not light up with these connections. If it does, the triac is leaky (FIG. 4-27). Push down on the momentary switch and rotate R2 until the red light remains on—no matter where the control is adjusted. Let up on SW1.

4-27 A normal light test until SW1 is pushed. The light will stay on when SW1 is released, if the triac is normal.

With the normal triac, the red light will remain bright when the gate is triggered on. Now, remove the green gate clip and the red (LED) remains on with a normal triac (FIG. 4-28). If the light goes out, suspect a defective triac.

4-28 The different leakage, open, and normal tests with the triac checker connected to the suspected triac.

Use only the red and black terminal leads to check for a leaky triac. The red light is on and normal when the leads are attached to the MT1 and gate (3) terminals. Does the LED light with reverse leads between black and green terminals? If the LED lights when the red and black leads are connected to the gate and MT2 the triac is shorted or leaky. The defective triac will show leakage between MT1 and MT2. Also, this little tester may be used as a continuity tester by using the red and black clip leads.

Parts list for triac checker

T1	12.6 Vac 450 ma power transformer
C1	1,000 35 V electrocapacitor
D1	1 amp silicon diode
R1	270 $1/2$ watt resistor
R2	1k $1/2$ watt resistor
R3	50k linear control
SW1	Momentary spst switch
LED	Red
Misc.	Hookup wire, three different colored alligator clips, plastic box, etc.

Chapter 5

Troubleshooting microwave oven switches

More problems are caused by a defective interlock switch than any other component within the oven. Perhaps it's because several interlock, defect, monitor, and safety switches are activated when the door is constantly opened and closed (FIG. 5-1). These switches are located at the top and bottom of the door area and some directly behind the door handle. Primarily, the many interlock switches are installed to protect the operator from injury and exposure to rf radiation. The interlock switches are used to shut off the oven should the door be opened while the oven is in operation.

The small interlock switch has a microswitching action with only a light touch of the small plastic plunger. These switches come in normally open and normally closed positions. Look closely, they are all marked on the side of the switch area. The switch may be an spst or spdt type. A three-prong or terminal interlock switch is an spdt type. Usually, the common connection is located at the bottom terminal of the switch.

These interlock switches have different voltage and current ratings. The voltage reading may be the same, but different current ratings are marked on the body of the microswitch. Always replace the interlock switch with one having the very same current rating or one with a larger current rating. These switches are all about the same size. They may be interchanged from one oven to the next as long as the current rating is correct with a normally closed or normally open interlock switch. You may find one or two ovens with larger interlock switches that are not interchangeable since they will not mount in the same area.

The microswitch may become defective with poor or open internal contact points. Dirty or pitted contacts may let the oven operate intermit-

5-1   The location of the upper and lower interlock switches might be right behind the open door. Some interlock switches are located alongside the oven wall (courtesy of Norelco Corp.).

tently in an erratic manner. The switch contact spring may become broken causing the small plunger to rest even with the surface switch area. Suspect a defective microswitch when an spdt switch may make contact in one position and be open in the other.

A defective interlock switch may be locked in one position and may not release. Sometimes, the defective interlock may be located by the appearance of the switch terminals. Check the interlock switch terminal for burned marks. When the plastic insulation is curled back or appears black, suspect a poor terminal connection or poor internal switch points (FIG. 5-2). Suspect food or liquid alongside the small plunger locking it in the downward position. Replace both the switch and connection when this condition occurs. You may solder the old connection to the new switch terminal for good contact. Otherwise, the new switch may be destroyed by a poor terminal connection and erratic oven operation.

All interlock switches may be easily checked with the low ohmmeter range of the vom. Make sure the power cord is pulled out and the high-voltage (hv) capacitor discharged before attempting any ohmmeter measurements. Set the ohmmeter to R×1 and connect both leads across the

5-2 Notice the black appearance of the middle interlock switch terminal. Either the switch has poor contacts intermittently or poor clamp-on connections. Replace both the switch and the connection.

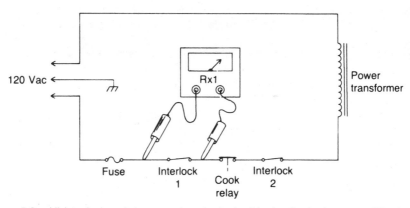

5-3 All interlock switches may be checked with the R×1 ohmmeter. Clip the leads across the suspected interlock switch and open the oven door. The ohmmeter hand should show a short and then open according to the type of interlock switch.

switch terminals (FIG. 5-3). It's best to have alligator clips on both vom terminal leads so they can be clipped to the switch terminals. In some cases, you may have to remove one terminal to make a good connection. Always write down the color codes of the wire terminals when three wires are connected to the switch. It is not necessary to record the connection of a two-terminal type interlock switch.

Now, open the door and notice the meter movement of the vom. A NO switch will not show a reading while the NC switch will have a shorted measurement. If any type of resistance is noticed between the two terminals with an NC switch, suspect dirty contacts. Replace the switch at once. A good switch will open and close when the door is opened and closed in their respective switching modes.

You may find several names for the many micro-interlock switches. Some ovens list the interlock switches as primary and secondary interlock switches, while others may call them latch or defeat switches. Besides being called the monitor switch, some ovens may call it a safety switch. Although the names may be different, the small microswitches perform the same task in the microwave ovens. You may also locate upper and lower interlock switches in some ovens. You may find one or two interlock switches ganged with an oven lamp switch.

Before leaving the microwave oven, be sure to check for interlock action. Simply open the door and listen for the switches to make or break contact. The door should be adjusted for a very tight fit with interlock action. Never bypass or remove an interlock switch. Always replace the defective switch before attempting to operate the oven. Often, the oven will not operate with a defective interlock switch.

INTERLOCK ACTION

The interlock switches may be located with the primary and secondary circuits. You may locate them in both sides of the power line or only one side going to the power transformer (FIG. 5-4). In early Sharp models, the interlock monitor and fuse are mounted on one replaceable assembly (FIG. 5-5). The interlock switches are so placed within the circuit so no matter what happens, the oven will not operate with the door open.

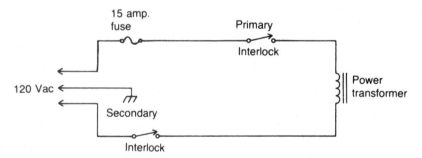

5-4 The interlock switches may be located on both sides of the power-line circuit. These switch contacts control the power-line voltage to the power transformer for proper oven cooking.

Even in a simple microwave oven when the door is closed all the door interlock switches are activated. Either the switches are located right behind the door or they may be tripped by a metal lever at the bottom of the door and the interlock switch assembly located on one side of the

5-5 In some Sharp ovens, the monitor and interlock switches are located on a separate replacement assembly with the 15 amp fuse. Replace the whole assembly when the monitor switch keeps opening up the fuse.

oven cabinet. The primary and secondary switches are activated, when the door is closed, by the contacts of the monitor or safety switch opening up.

A latch or interlock switch-hook assembly is mounted on the door and when closed it trips the small microswitches. Some of these switch-hook or trip assemblies are movable or flexible while others are mounted stationary on the door. A warped or out of line door may not let the hook assembly strike the monitor switch assembly resulting in no interlock switch action. So don't overlook the most obvious problems, such as a broken hook spring, misaligned hook assembly, and extended switch assembly.

You may locate a defective oven operation by taking interlock switch testing procedures or checking the door alignment. Most ovens have an interlock switch testing procedure using an ohmmeter. In case the service literature is not handy for a given oven, take ohmmeter tests across the suspected interlock switch. Remember, the interlock switch controls are closed when the door is closed, except for the monitor switch. Suspect a defective interlock switch when you have no reading across the switch terminals. The following is a typical oven interlock switch testing procedure.

Checking the left interlock switch

1. Remove the power plug.
2. Connect one lead of the ohmmeter to the blue lead at the bottom timer terminal.
3. Connect the other ohmmeter lead to the orange wire of the timer switch.
4. Firmly close the door. The meter hand should show a direct short. Replace the interlock switch if the reading is erratic or if there is no reading.
5. Open the door slowly and when the door is about $1/4$ inch open, the meter hand should show an open circuit. If the meter hand does not show an open circuit, check for correct interlock switch adjustment.

Checking the right interlock switch

1. Remove the power cord and close the door.
2. Attach one lead of the ohmmeter to the terminal board upon the bulkhead.
3. Attach the other meter lead to the white wire on the surge relay or the fuse.
4. Open and close the door. The meter should show continuity when the door is closed. The switch should open the circuit before the door exceeds $7/16$ of an inch.
5. If the switch doesn't close the circuit or if it is erratic in operation, suspect an open or dirty interlock switch contact. Check the operation of all interlock switches after completing the service of the microwave oven.

Norelco Model S7500 interlock switches   The latch (primary) protective switch is activated by the latch when it is closed. The switch is in an open position (when the door latch is open) and interrupts the current to the magnetron and probe circuit. Closing the door latch will close the latch (primary) switch, and normal operations can be continued.

The door (secondary) switch is located behind the front frame and is activated when the door is closed. When the door is open, it will act as a safeguard to interrupt the circuit if, for some reason, the latch (primary) switch should cease to function properly and remains closed with the door open (FIG. 5-6).

The sense interlock switch disables the control panel if the door is open or the switch is out of adjustment. This switch is $1/2$ of the secondary switch. It is considered a low-level interlock, meaning that a full 115 V does not flow through it as it does in the regular secondary interlock.

The interlock sense switch is normally closed. If the lead is put on the normally open terminal, the control panel can be operated with the door open, but the magnetron will not power up under this condition.

(Normally closed)
safety switch

(Normally open)
door switch

(Normally open)
latch switch

Interlock
sense switch

(normally closed)

5-6 The interlock switch assembly in the Norelco Model S7500 microwave oven (courtesy of Norelco Corp.).

The safety switch, which is normally closed, along with the latch switch and door switch (both normally open), is activated by closing the door. As the door is closed, the plunger or safety switch is pushed and the switch is opened (FIG. 5-7). As the door continues to close, the pin on the door closes the door switch. When the latch is closed, the safety switch is closed and the circuit is complete.

If, for some reason, the latch switch and secondary switch stick in a closed position and you tried to open the door, the circuit would be shorted through the normally closed safety switch. This would blow the fuse. In some ovens, the safety switch is called the monitor or short switch.

If a switch assembly is replaced, it must be adjusted. If, at any time, the oven fuse blows due to an oven fault or a circuit breaker trips, often from being reset one time with the house circuit unloaded except the oven, the three switches must be replaced as an assembly. This is because the short circuit current will have damaged these important switches. Throw the defective switch assembly away.

REPLACING A DEFECTIVE INTERLOCK SWITCH

After locating a defective interlock switch with the ohmmeter, remove and replace it. Although some of the latch or interlock switches are out in

5-7 The plunger or actuator in the door of the microwave oven engages the interlock switch. Notice if the plunger strikes the interlock or if it slides by.

the open, you may encounter some that are difficult to remove. Besides removing components around the switch area, you may have to loosen up or pull out the front control assembly. Most manufacturers list the different components that must be removed to get at the interlock switches. With no available service literature, check the area over carefully before removing any other components.

The defective interlock switch should be replaced with the original manufacturer's part number. When an interlock or latch switch is not readily available, you may be able to use another oven microswitch instead. Make sure the switch current and voltage is the same with proper normally closed or normally open operation. You may find the defective switch riveted to a metal switch assembly. Simply grind off the metal rivet. Mount the new switch with small bolts and nuts.

You may find that some interlock switches just clip into a plastic switch assembly or are held in place with one metal screw. These interlock switches are easily exchanged with other oven switches. Do not replace a smaller type interlock switch with a large one. Remember, the switch must mount so the door latch will trip the small microswitch plunger. When available, always replace with original parts for safety and easy replacement. Realign the switch and door assembly.

INTERLOCK SWITCH REPLACEMENT

Proper adjustment of the interlock switch assembly may also help to adjust the door for any play between the door and oven. After replacing the interlock switch, check the switch assembly for proper latch head alignment. The latch or interlock switch head must trip the microswitch with no play of the door. Usually, the interlock switch assembly can be moved back and forth with metal screws. Pull the assembly back away from the door with hand pressure against the door. Now, tighten the screws of the interlock switch assembly. Check for play in the door by pulling on the door.

Especially check the bottom interlock door switch assembly. Do not press the door release button while making any adjustments. Make sure the latch or interlock key or hook assembly moves smoothly after each adjustment is completed and the mounting screws are tight. Sometimes a dab of cement or glue on the screw heads or nuts will keep the interlock switch assembly from loosening up. Double check for smooth operation of the upper latch and bottom latch head when activated by one long bar assembly. Improper latch head action may be caused by a binding lever or missing spring of the latch head assembly. Follow the manufacturer's adjustment procedure.

Here is a list of various microwave oven problems that were caused by defective interlock switches.

Dead

1. Replaced broken door latch head, spring and interlock switch. Installed new 15 amp fuse and adjusted door and switch assembly.
2. Installed new latch switch assembly.
3. Replaced 15 amp fuse with new monitor switch assembly.
4. Defective top interlock switch.
5. Replaced interlock switch assembly and realigned door.

Intermittent and erratic operation

1. Replaced upper interlock switch assembly.
2. Door latch assembly loose (check pop rivets).
3. Realigned door as latch head would sometimes not engage interlock microswitch.
4. Installed new door spring at the top (would not operate door switch).
5. Replaced latch switch and excessive door play.
6. Replaced defective interlock switch assembly.

Oven light on—no cooking

1. Dead when pushed cook button (replaced interlock switch).
2. Lights up but no fan or cooking (adjusted door and interlock switch assembly).
3. Have to pull up door to make oven cook (realignment of door and adjusted switch assembly).

4. Press in on door to cook (realignment of interlock and monitor switch assembly on side of oven cavity).

Panasonic safety switch system While there may be a slight difference in construction, most microwave ovens have three safety switches, namely the *primary*, *secondary*, and *safety* switches. These switches are activated by the movement of the door. The primary and secondary interlock switches shut off the power supply to the appliance when the door starts to open. The safety switch monitors both switches and blows the line fuse to de-energize the appliance in case both interlock switches malfunction and do not turn the power off (FIG. 5-8).

5-8 The safety switch in this Panasonic oven is located in about the same area as most ovens (courtesy of Matsushita Electric Corporation of America).

The basic function of the primary and secondary interlock switches is to interrupt the power supply to the magnetron, thereby stopping the microwave oscillation when the door starts to open during operation. The safety switch, which is wired to close the circuit when the door is opened, creates a short circuit when both or either interlock switches (depending upon the applicable regulation) fail to operate and causes the line fuse to blow and shut down the operation of the appliance. There is a time lag between the operation of the primary and secondary interlocks and the safety switch so that the line fuse remains intact during normal operations of both interlock switches (FIG. 5-9).

Panasonic's interlock switch adjustments When the 15 amp fuse is blown, due to operation of short switch, you must replace latch switch A and the short switch. For USA models, also replace power relay B when

5-9 The safety switch in the Panasonic ovens may be located in several different areas in the different models (courtesy of Matsushita Electric Corporation of America).

the continuity check reads shorted contacts 3 and 4. In replacing faulty switches, be sure mounting tabs are not bent, broken, or otherwise deficient in their ability to hold the switches (FIG. 5-10).

1. When mounting latch switch A, latch switch B, and short switch to door hook assembly, mount the latch switch A, the latch switch B, and the short switch to the door assembly (FIG. 5-10). No specific adjustment is required of these three switches.

5-10 After mounting new switches, confirm that the gap between the switch housing and the switch actuator lever is no more than 1.0mm when the door is closed (courtesy of Matsushita Electric Corporation of America).

2. When mounting the door hook assembly, adjust the door hook assembly by moving it in the direction of arrow (FIG. 5-11), so that the oven door will not have any play in it. Check for play in the door by pulling the door assembly. Make sure that the latch keys move smoothly after adjustment is completed. Completely tighten the screws holding the door hook assembly to the oven assembly.

Screws

Door hook assembly

5-11 Check for play in the door and when mounting the door hook assembly. Adjust the door assembly by moving it in the direction of the arrows (courtesy of Matsushita Electric Corporation of America).

THE MONITOR SWITCH

The monitor or safety switch is intended to render the oven inoperative by means of blowing the 15 amp fuse when the contacts of the interlock switch fails to open when the door is opened. In case the oven keeps on operating with the door open, the monitor switch blows the fuse, shutting the oven down to prevent radiation burns or injury to the operator. The monitor switch contacts are always open when the door is closed. Since the 15 amp fuse is in one side of the power line, the monitor switch automatically blows the fuse (FIG. 5-12).

Look at FIG. 5-13. If the contacts of interlock switch 1 and the contacts of latch switch 2 malfunction (or only interlock switch 1 remains closed) the 15 amp fuse will open due to the large current surge covered by the short or monitor switch. The magnetron tube stops oscillating.

5-12 The monitor or safety switch automatically blows the fuse when latch-switch I remains closed. A monitor, safety, or short switch is intended to render the oven inoperative by blowing the fuse when the door is opened.

5-13 The monitor or safety switch blows the fuse when interlock-switch I remains closed, shutting down the magnetron tube. Always replace both safety and interlock switches since the heavy power-line current is fed through them. Actually, the power-line voltage is applied directly across the switches and fuse.

Always replace both latch switch 1 and the safety or monitor switch when the fuse keeps opening. Full power-line voltage is placed across the contacts of each switch before the fuse opens. In some models, the whole monitor-interlock fuse assembly is replaced when the shorting condition occurs.

In case the fuse blows after opening the front door, you may assume the interlock switch contacts are closed or arced over and will not open up. Sometimes the fuse will open with too much play in the front door. This may let the interlock switch stay closed with the monitor or safety switch closing, thus knocking out the fuse. The interlock switch assembly should be replaced and properly adjusted so the door has no play between the door and cabinet. You may find the interlock switch arm assembly is hanging up, keeping the interlock switch contacts closed, causing the fuse to blow. Sometimes the interlock switch plunger contacts are held down by liquid or food spilled down inside the switch area. Always replace both the interlock and monitor switch. Clip the test light and socket across the fuse holder so that you don't keep blowing fuses while locating the defective safety switch. The light will be bright with a shorted interlock switch.

The monitor or safety switches can be checked with the R×1 range of the ohmmeter. Simply place the meter leads or clips across the suspected safety switch. When the door is closed, the switch contacts are normally open. If not, the safety switch is defective. Now, open the oven door. The switch contacts will close with the door open. Check the service literature for the location of the monitor or safety switch.

In early Sharp microwave ovens, a monitor interlock switch assembly is located in the left side of the oven (FIG. 5-14). A long door lever-bar triggers or operates the interlock assembly. The 15 amp fuse, interlock switch, and monitor switch are located on the same metal switch assembly. Simply adjust the whole monitor switch assembly by loosening two nuts. Usually, the oven cannot be turned on since the door lever may not trigger the interlock switch. Snug up the mounting bolts very tight. Place a dab of glue over each nut so the monitor switch assembly will not loosen up.

5-14 In a Sharp R7600 model, the interlock and monitor switches are located on the left side of the oven. The back cover must be removed to get at this switch assembly. Adjustment of this switch is necessary to make the oven operate.

Some monitor or safety switches are in a fixed position so no adjustment is needed, while the secondary interlock monitor switch assembly of other ovens must be adjusted for correct operation by inserting this spacer material (.004 inch or .008 inch) between switch plunger and stopper to energize the secondary interlock switch before the monitor switch is activated. If it is not, when the door is opened, the secondary switch will not open before the monitor switch closes, thus opening up the 15 amp fuse. Most safety switch adjustments are made by loosening the two mounting screws or metal screws holding the whole switch assembly. Slide the switch assembly until the lever trips the microswitch and then tighten down the mounting nuts. Follow the manufacturer's monitor switch adjustment procedure for correct oven operation.

Norelco S7500 safety switch adjustment Loosen the two screws that secure the switch on top until the assembly can be moved freely back and forth (FIG. 5-15). Close the door and secure the latch. Remove the wire from one side of the switch and place the leads of the ohmmeter across the terminals of the switch. Move the switch toward the door until it is lightly stopped (switch is normally closed). Check by opening and closing the door. As the door is opened, the meter should show an open circuit. Tighten and secure the two switch screws and replace the wire.

5-15 Loosen the two screws so that the safety switch can be moved freely. Check the safety switch terminals with the ohmmeter leads clipped across the switch terminals (courtesy of Norelco Corp.).

THE START AND STOP SWITCH

In older models, the oven was turned on by a heavy duty type off/on switch. Since the switch supplied the entire oven current load, these switches had a tendency to arc internally and after several years must be replaced. They would arc and sputter sometimes making poor contact, then again the oven was inoperative. These switches must be replaced with the original part number.

You may find a separate start and stop switch in ovens with a panel controller assembly (FIG. 5-16). The start switch is pushed to start up the control assembly while the stop switch is a closed switch. The start switch is normally open until punched. When the stop button is pushed, the switch disengages the control assembly from the circuit.

5-16 You may find separate start and stop switches in an oven with a controller assembly. The start switch is momentarily closed when pushed, while the stop switch is in a fixed close position.

Both switches may be checked with the R×1 scale of the ohmmeter. Clip the meter leads across the switch terminals. Always pull the power plug and discharge the hv capacitor when taking ohmmeter readings. Push in on the start switch. The meter should read a direct short or less than 1 ohm with the start button pushed in. Do likewise with the stop switch. A directly shorted reading should be obtained with the meter leads clipped across the switch terminals. Now, push in on the stop switch. The meter should show an open circuit.

The oven or cook switch is normally open and when pushed will make contact energizing the oven relay or triac assembly, applying line voltage to the power transformer (FIG. 5-17). In some lower priced or commercial ovens, the cook switch may be in series with one side of the power line, connecting voltage directly to the power transformer (FIG. 5-18). You may find another rotary type switch that provides voltage to the defrost circuits or cooking modes. Actually, the defrost switch contacts are in parallel with the cook switch except that in the defrost position the switch is controlled off and on by the defrost timer assembly.

5-17 The oven or cook switch is normally open and, when pushed, makes contact energizing the oven relay or triac assembly. You may find a large separate oven cook switch in other ovens.

5-18 In commercial ovens, the cook switch is a heavy-duty switch located in one side of the power-line circuit. These switches have a tendency to pit and arc after several years of operation. They must be replaced with the original type of cook switch.

Like all switches, the oven or cook switch may be checked with the ohmmeter. Remember, the cook switch that supplies voltage to the oven relay only makes contact while pushed in. When you let up on the cook switch, the oven or cook relay contacts take over and are in parallel with the cook switch contacts. Suspect defective oven relay contacts if the oven will only operate by holding in on the cook button (FIG. 5-19). Either a wire is broken off or the relay contacts are defective.

5-19 Suspect defective oven hold contacts of the oven relay if the oven will only operate by holding the cook button. This set of contacts is in parallel with the oven cook button.

Norelco S7500 on/off switch tests Remove the on/off switch and wires and tag them for identification. Set the multimeter on R×1 ohm scale and the meter on 0. Place one test lead on terminal 7 and the other on terminal 8. Depress the "on" push button. The meter should defect to the 0 mark. With the leads on the same terminals, depress the "off" button. The meter should indicate an infinite amount of resistance. Remove the leads and place one on terminal 3 and the other on terminal 16 (FIG. 5-20).

Depress the "on" push button and watch the multimeter. When the button is pushed, the meter will go to 0. Push the "off" button and the meter should indicate an open circuit. Place the leads on terminal 1 and 17. Depress the "on" push button and watch the multimeter or DMM. When the button is pushed, the meter will go to 0. Push the "off" button. The meter should read an open circuit. If the meter hand changes and appears erratic, the switch terminals may be dirty, burned, or pitted.

THE OVEN RELAY

Line voltage is applied to the primary winding of the hv transformer by the cook switch, timer controlling switch, oven relay, or triac assembly (FIG. 5-21). Some ovens may call the relay a power relay, as power is

5-20 Checking the on/off switch in a Litton microwave oven with the ohmmeter connected to both sides of switch terminals.

5-21 Line voltage is applied to the power transformer by either a cook switch, timer switch, triac assembly or oven relay. You usually can hear the oven relay close by listening closely.

applied to the power transformer. Often, you may hear the oven relay energize.

The cook or power relay may be checked with a vom. With the power off, locate the solenoid coil winding terminal and check for continuity (FIG. 5-22). Most oven relays will have a resistance between 150 and 300 ohms. If in doubt, remove the external cable leads from the solenoid terminals. Then take another resistance measurement. Of course, if the coil is open, the relay must be replaced.

5-22 Remove the power cord and discharge the hv capacitor before measuring continuity of the solenoid. Most oven relays have a resistance between 150 and 650 ohms.

A continuity check between switch contact terminals should indicate an infinite resistance. Only when the coil is energized should you measure a short across the switch terminals. A continuity check between each switch terminal and ground should indicate an infinite resistance. The switch terminals may be checked by removing the top cover and using an insulating tool to push the solenoid together.

In case voltage is not reaching the solenoid terminals, the oven relay will not operate. Some oven relays operate from the ac power line, while those with a control board assembly operate from 10 to 24 Vdc. The voltage may be checked at the solenoid terminals when the oven will not turn on (FIG. 5-23). Always clip the meter leads to these terminals. First, determine if the voltage is ac or dc. Low or no voltage at the relay indicates a

5-23 For stubborn cases, clip the voltmeter to the solenoid winding of the relay to determine if the circuit voltage is getting to the relay. Some ovens use the power-line voltage across the relay terminals, while ovens with control board assemblies have a voltage from 14 to 35 Vdc.

defective control-board circuit. Correct voltage at the solenoid terminal indicates an open or defective relay solenoid.

When everything else has been checked and the relay will not energize, remove the solenoid terminal wires. Mark where each wire goes before removing them. Now apply 14 to 20 Vdc from a bench power supply to the low-voltage relay. A normal relay will operate the cooking cycle. With a suspected open solenoid winding, the contacts may be checked by pushing down on the contact point assembly with an insulated tool with the oven in operation. Use this method as a last resort, being careful not to touch any other components.

The relay contact voltage may be measured at the primary winding of the hv power transformer (120 Vac). Clip a vom across the primary leads of the transformer to monitor the applied voltage. When the oven relay operates normally, the line voltage is measured with the voltmeter. Besides a defective oven relay, you may encounter a leaky triac assembly.

RELAY CONTACT LAYOUT

The oven relay may have two or three different circuits, switched in or out, and a solenoid winding. The oven relay winding may be controlled by the ac power-line voltage or the dc voltage from the control circuits (FIG. 5-24). The contacts and winding spade terminals may come out at the top of the relay. Locate the relay winding by visual means or with the ohmmeter. The resistance across the oven relay winding may be from 150 to 650 ohms. Check each set of switching contacts with the ohmmeter

5-24 The relay contact spade connection may be found at the top of the unit. Check the relay points and the coil winding with the ohmmeter. Energize the switch contacts by holding down the relay points and plunger.

across them by manually depressing the oven switch to locate poor or open contacts.

In some ovens you may encounter another auxiliary relay (FIG. 5-25). Here, the auxiliary relay turns on the stirrer and blower motor. The relay is controlled from the electronic controller assembly. Notice the resistance of this particular relay has a total of 1.3 kilohms. All relays should be replaced with the original part number.

Panasonic's power relay control There are basically two types of magnetron duty control systems available in the field:

- To control the on/off time of magnetron in the primary circuits.
- To control the on/off time of magnetron in the secondary circuits.

5-25   An auxiliary relay may be used in some ovens to control the stirrer and blower motor. The relay voltage is controlled by the electric control board.

The primary duty cycle circuit controls the duty cycle of the magnetron in the primary circuit (FIG. 5-26). The primary power supply to the hv transformer is interrupted intermittently by means of a control circuitry and a switching device. The switching device may be either a power relay or triac. The control circuitry controls the interrupting periods of the magnetron within a specified duty cycle and varies the average power output of the microwave oven. Since this system controls the circuit in the primary circuit, the switching device requires a lower level of insulation; however, a higher stress to the components in the secondary circuit would result due to transient currents causing a possible shortening of the component's life.

The secondary duty control circuit controls the duty cycle of the magnetron in the secondary circuit (FIG. 5-27). The high voltage to the magnetron is interrupted intermittently by means of an associated control circuitry and a switching device. The control system of the magnetron duty cycle is equivalent to the primary duty cycle control circuit, except that the switching device is in the secondary circuit.

The advantage of this system is that it results in less stress to the components in the hv circuit since it switches the circuit when no high voltage is being applied so no transient currents are generated. The switching device, however, requires a higher grade of insulation because the device is located in the hv circuit.

When the power selection is set at any power setting other than high power, the variable power switch (switching device) is energized intermit-

5-26 The power control of this Panasonic oven is located in the primary circuits with the switching device or power relay (courtesy of Matsushita Electric Corporation of America).

tently by signals from the variable power control circuit. The variable power control circuit controls the on/off time of the variable power switch contacts within a specified duty cycle. With most Panasonic ovens, one complete duty cycle is approximately 22 seconds (FIG. 5-28).

Assuming the full power of the microwave oven is 700 watts, the average output power at medium power setting is approximately 477 watts. If the full power of the microwave oven is 600 watts, the average output power of medium power setting is approximately 409 watts. In this way, the output power of the microwave oven be varied to any desired power setting by changing the duty cycle ratio.

5-27 The secondary duty control circuits for the duty cycle magnetron in this Panasonic oven is controlling the high-voltage to the magnetron tube (courtesy of Matsushita Electric Corporation of America).

INDICATION	OUTPUT POWER AGAINST HIGH POWER	ON. OFF TIME ON VARIABLE POWER SWITCH		
High	22 / 22 (100%)	⊢— 22 (S) —⊣⊢— 22 (S) —⊣ ON / ON		
Med	15 / 22 (68%)	⊢— 15 (S) —⊣7 (S)⊢ ON / OFF / ON / OFF		
Low	9 / 22 (41%)	⊢9 (S)⊣⊢13 (S)⊣ ON / OFF / ON / OFF		
Warm	2 / 22 (10%)	2 (S) ⊢— 20 (S) —⊣ ON / OFF / ON / OFF		
Defrost	10 / 22 (45%)	⊢10 (S)⊣⊢12 (S)⊣ ON / OFF / ON / OFF		

5-28 The medium indication of the Panasonic variable power operation is on 15 seconds and off 5 seconds with a 68% output power against high power (courtesy of Matsushita Electric Corporation of America).

The coil of the variable power switch is energized intermittently by the digital programmer circuit in some microwave ovens (FIG. 5-29). The DPC controls the on/off time of the variable power switch contacts to vary the average output power of the microwave oven from warm to high power. The defrost power and time is selected with the start pad of the programmer circuit. Notice the variable power switch or relay is located in the hv circuit of the magnetron (FIG. 5-30). The on and off cooking time controls the high voltage at intermittent intervals to the magnetron tube.

5-29 The coil of a variable power switch is controlled by the DPC in some ovens. In turn, the variable switch turns the magnetron on and off.

TIMER CONTROLS

You may find one or two timer control assemblies in the manual or commercial microwave ovens (FIG. 5-31). The top timer may have a scale from 0 to 5 minutes, while the second timer has a longer cooking time. Set the

5-30 Notice the variable power switch or relay is located in the hv circuit of the magnetron. The electric controller controls the time of the magnetron oscillations.

5-31 You might find one or two timer controls assemblies in the manual commercial microwave oven. The top timer might have a scale from zero to five minutes, while the second timer requires a longer cooking time.

timer dial for the desired cooking time and the blower motor will rotate. The oven light will come on at once (FIG. 5-32). Actually, when the correct cooking time is set, the timer switch contacts close in the timer assembly. In this particular circuit, notice the timer will not start to operate until the cook switch is pushed, as the timer motor connects into the ac circuit after the oven or cook switch.

The timer switch contacts may be checked with the ohmmeter. Pull the power plug and discharge the hv capacitor. Clip the meter leads across the timer switch terminals. Now rotate or set the timer for 3 or 4 minutes. The meter hand should indicate short or continuity if the switch is normal. An erratic ohmmeter reading may indicate a dirty switch contact. Clean or spray the switch with cleaning fluid. If the switch remains intermittent, replace the timer assembly.

5-32 Usually, when the timer is set, the blower motor and the oven light will come on. Notice that in this circuit, the timer does not start until the cook switch is depressed.

A defective timer may become intermittent in operation or not shut off. If the timer motor fails to run, check for line voltage (120 Vac) across the motor terminals. The timer motor should rotate when the cook button is pushed. In case the motor does not operate with the correct ac voltage applied to the timer motor terminals, suspect jammed gears or an open motor coil. Remove the one terminal from the motor coil and check for continuity with the ohmmeter. Replace the timer assembly if the motor is open or is erratic in operation.

OVEN LAMP SWITCH

Although most oven bulbs are turned on when the door is opened, you may find a separate lamp switch (FIG. 5-33). In this particular circuit, the lamp is turned on when the door is opened and when the power relay is energized. All of this occurs with a small microswitch ganged with the latch and short switch.

You may locate one or two oven lamps in some ovens. First, replace the lamp bulb before tearing into the oven. If the oven lamp does not come on, check for a defective lamp switch. Suspect a defective lamp socket or connecting wire after replacing the bulb and checking the lamp switch. You should measure the entire line voltage across the lamp socket terminals (120 Vac).

The same oven lamp may serve as a cook lamp in FIG. 5-33, or a separate neon type indicator may be found (FIG. 5-34). Here, the neon bulb and voltage dropping resistor are included in one component mounted on the front panel. When the cook lamp comes on, you know line voltage is applied to the primary winding of the hv transformer.

The oven lamp may be located in the light switch circuits or after the oven relay is energized. In some ovens, when the light switch is turned on, the oven lamp is on. When the oven is operating, the oven lamp is turned on with the control relay. In the Norelco RR7000 model, the oven lamp is controlled entirely by the control panel circuit (FIG. 5-35).

5-33 The oven light may come on when the door is opened or closed with some different oven switches. Here an spst switch is used for this purpose.

5-34 A neon bulb with a voltage-dropping resistor indicates when the cook relay is operating. This light could also indicate that the voltage has reached this far in the circuit.

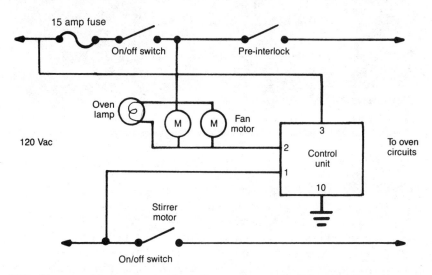

5-35 The oven light is operated from the panel control unit in a Norelco microwave oven (courtesy of Norelco Corp.).

5-36 Another switch, called the select switch, turns on the defrost circuit and cook operation. Notice that the separate defrost switch is turned on when the defrost motor is operating, turning the cook process on and off in the defrost mode.

5-37 The cook and defrost switches are locked together in a Litton microwave oven. The cook and defrost switch intermittently applies power to the magnetron tube.

THE SELECT SWITCH

The select switch is only found in some oven circuits to select cooking or defrost modes. Usually, the select switch is paralleled with the defrost switch. Check the select switch for dirty or defective contacts if the switch operates in select mode and not in defrost. When the select switch is rotated to defrost, the defrost motor starts up, closing the defrost switch across the select switch contacts (FIG. 5-36). No defrost motor operation may indicate a defective motor or select switch (FIG. 5-37).

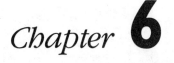

Chapter **6**

Servicing low-voltage problems

All low-voltage problems with microwave ovens are related to the power-line voltage. When low voltage or no voltage is found at the primary winding of the large power transformer, check for low-voltage problems (FIG. 6-1).

In other words, a defective component is preventing ac voltage from being applied to the primary winding of the power transformer. The defective component may be located in either side of the power line. Any part from the power cord to the transformer winding may be the defective component (FIG. 6-2).

LOW-VOLTAGE PROCEDURES

To determine what component may be defective, use existing lights and fan motor as indicators. You will find many lamps and motors throughout the low-voltage circuits. No light or fan motor rotation may indicate an open fuse, a defective power cord, or defective switch. When the oven lamp is on the fuse is okay.

Usually, when the oven lamp is on we know that the power-line voltage is normal to the oven relay. This may indicate that all interlock switches, thermal cutout, and timer circuits are normal. If the oven relay does not come on, you may locate a defective cook switch, cook relay, or poor contacts on the oven relay.

In ovens with digital programmer control (DPC) circuits, a defective control board, thermal cavity fuse, thermal protection, triac, or power relay may prevent line voltage applied to the primary winding of the power transformer. First check all fuses and triac with the ohmmeter tests. Next check the power relay components and circuits before replacing or

6-1 The interlock switches, fuse, oven relay, or triac are in the power-line of the low-voltage circuits.

6-2 Measure the power-line voltage (120 Vac) at the primary winding of the power transformer to determine if a component is defective in the low-voltage circuit. Any part between the power cord and the transformer may be the defective component.

checking out the control board. Servicing the control board and replacement data is given in chapter 10.

A normally lighted cook lamp indicates power-line voltage is applied past the oven relay assembly (FIG. 6-3). Remember, in some ovens the control programmer may turn on the oven cavity light. Check the defrost

6-3 A normally lighted cook lamp indicates the power-line voltage is applied after the oven relay assembly. Use the lamp and motor operation to help signal trace the low voltage circuit.

lamp and defrost motor for a low-voltage indicator in manually operated ovens. The turntable motor rotation of a Sharp oven is another component that will indicate if the power voltage is applied up to the primary winding of the power transformer. Usually the fan blower will come on, indicating low-voltage should be applied to the transformer. In a Quasar oven, the rotation of the fan blower indicates voltage is applied to the power-transformer circuits (FIG. 6-4). By simply checking the light bulbs and various motor operations, you may quickly isolate the defective component in the low-voltage circuits.

CHECKING FUSES

Besides the 15 or 20 amp line-voltage fuse, you may find a thermal cavity fuse in series with the line fuse, power relay, and thermal protector (FIG. 6-5). All fuses can be quickly checked with the R×1 scale of the ohmmeter. Always replace the exact type fuse and amperage in each oven. Inspect the fuse holder for poor connections. Sometimes the fuse clips are spread and may not fit tightly around the end of the fuse causing intermittent oven operation. The thermal fuse is enclosed in a metal case.

KEEPS BLOWING FUSES

Besides the interlock switch, the 15 amp fuse is replaced more often than any other component in the microwave oven. Most ovens use a 15 amp chemical cartridge fuse (FIG. 6-6). The fuse may be located at the top or bottom of the oven area. You can't go wrong as only one fuse is found in

6-4   In a Quasar oven, the rotation of the fan blower indicates that voltage is applied to the power transformer circuit. The fan motor is paralleled with the transformer winding.

6-5   There is both a 20 amp and a terminal-cavity fuse in this low-voltage circuit.

any microwave oven. All oven distributors and manufacturers supply the 15 amp replaceable fuse. These fuses may be obtained from the local electrical and hardware stores. Do not replace the fuse with one with a higher amperage rating.

The 15 amp fuse may open or blow when there is an overloaded condition, power-line outage, internal motor switch fails, or during lightning storms. Sometimes when the oven door is not adjusted properly, a monitor or interlock switch may hang up causing the fuse to open. The interlock switch locked in the on position may cause the fuse to blow when

6-6 Most ovens use a 15 or 20 amp fuse for protection. The small fuse may be located either at the top or bottom of the oven area.

the oven door is opened. Improper door alignment may cause the fuse to open.

Ask several important questions when the owner complains the fuse keeps blowing in the oven. Does the fuse blow when the door is opened or closed? Does the fuse blow after the oven has operated for several hours? Does the fuse blow when the power cord is plugged in or the oven is turned on? Does the fuse open when the door is moved? Has the fuse been replaced only once or several times? These questions may help to locate the defective component.

If the fuse blows when the door is opened, look for a defective interlock and monitor switch. In case the fuse blows when pulling up on the door, suspect a defective interlock switch. Check for a defective cord, plug or component on the power line when the fuse opens as the power cord is plugged in. If the fuse blows after operating several hours, suspect a high-voltage (hv) component that is operating quite hot or warm and breaks down under heat.

Check the fuse for open conditions with the R×1 ohm scale of the ohmmeter. You cannot see inside these chemical type fuses as you can with regular glass fuses. The open fuse will show no continuity reading. If the ends of the fuse appear loose, replace it. After replacing the open fuse, start the oven up. The oven may operate perfectly without any further service. When the fuse blows at once, inspect the oven for possible overloaded conditions.

Inspect the interlock and monitor switches for erratic operation (FIG. 6-7). Always replace both the interlock and monitor or safety switch when the fuse blows when opening the oven door. Replace the interlock switch when frozen or if it has poor switching contacts. Replace the monitor switch since this switch is placed directly across the power line with the 15 amp fuse (FIG. 6-8).

6-7   Inspect the interlock and monitor assembly for erratic operation. In this Quasar oven, the light switch is included with the safety and interlock switch assembly.

6-8   Always replace the monitor or safety switch because the switch is placed directly across the power line when the interlock switch hangs up. Naturally, the 15 amp fuse opens up.

Suspect a defective varistor or surge absorber in some ovens that keep blowing the fuse as the oven is turned on or plugged in. These small varistors are located within a plastic sleeve across the power line. They are included across the power line to prevent excessive damage to the oven components (FIG. 6-9). This same type varistor is found in the power-line circuits of TV receivers. So don't overlook one of these small varistors when the fuse continues to open (FIG. 6-10). When a power outage occurs or lightning strikes the power line, the varistor may be damaged and cause the fuse to open.

6-9   A varistor is placed across the power line to protect the oven from lightning and power-line surges. The varistor will arc over and blow the fuse.

6-10   Look for a burned varistor component when the fuse continues to blow as the power cord is plugged in. These varistors are like those found in most TV chassis.

VARISTOR TEST

If the varistor is suspected and may have a few black marks or missing areas, test it with the R×1,000 ohm scale. It's best to remove the varistor from the primary terminals of the controller transformer or ac power line. Inspect the varistor for damaged areas. Replace if damaged. Connect the ohmmeter leads across the varistor leads and a normal varistor should read infinite ohms. Replace if varistor has any type of resistance measurement.

In case the fuse keeps blowing, insert a 100 watt bulb across the fuse terminals. Remove the defective fuse and just clip the bulb leads to each fuse clip. You may save several dollars because these chemical fuses are not cheap. Simply leave the bulb in place until you have located the defective part. Sometimes, the light bulb may not indicate when a shorted component in the hv section is causing the fuse to open.

You may find an open fuse, replace it, and the oven begins to cook. Then in a few minutes, the cooking quits and you have a blown fuse. The magnetron may be leaky, open up the thermal cutout assembly. Although the magnetron may be the prime suspect, don't overlook a jammed blade in the cooking fan. If the fan blade doesn't rotate, the magnetron tube overheats. Likewise, the thermal protector switch shuts down the circuit if the fan motor is overheating, causing the fuse to blow. Don't overlook a defective fan motor when the fuse opens after several minutes of oven operation.

A shorted magnetron or leaky hv diode may cause the fuse to open. The fuse may blow after a few minutes or when the oven is shut off and this is sometimes caused by a leaky triac assembly. Check each component for possible damage after the oven has stopped working after an electrical storm. Usually, those components that tie into the power-line and the control-board assembly receive the most damage.

INTERMITTENT OVEN LAMP

The oven lamp may be used as an indicator to show that power is being applied to the oven and the fuse is good. The lamp may come on when the door is opened or closed. In some ovens, an spdt switch is used for this purpose (FIG. 6-11). Here the oven lamp is on with the door open so the operator may load the oven cavity. With the door closed, the oven lamp goes out until the power relay is in operation. When the lamp stays on with the door closed, replace the lamp switch, or suspect that a switch lever is hung up. Again, with this circuit, the oven lamp may be used as an indicator showing power is applied up to this point.

You may find two separate light bulbs in some microwave ovens. Often, the cavity lamp is 30 watts or smaller and may be replaced with a regular appliance-type bulb (FIG. 6-12). These bulbs are available from hardware, department, and electrical stores. All microwave oven distributors and manufacturers supply oven lamps for their particular ovens.

An erratic or intermittent oven lamp may be caused by a defective bulb, lamp, or interlock switch. You may even find the bulb is loose in its

6-11 The oven lamp is usually turned on by a switch when the door is opened or closed. An spdt switch is found in some ovens and is operated by the front door.

6-12 Often the cavity lamp is 40 watts or less. These regular appliance bulbs may be easily found at hardware, department, and electrical stores.

socket. Always replace the oven lamp while servicing the oven if the bulb appears black, even if it comes on. Usually, these bulbs do not last too long and this replacement may prevent a callback to replace just a lamp bulb.

Replace a suspected lamp switch when the bulb is intermittent or does not stay on. You may find the bulb is on with the door open, but will not come on with the oven operating. The oven lamp switch may have burned or dirty switch contacts. The cavity light switch may be found separately or ganged with other interlock switches (FIG. 6-13).

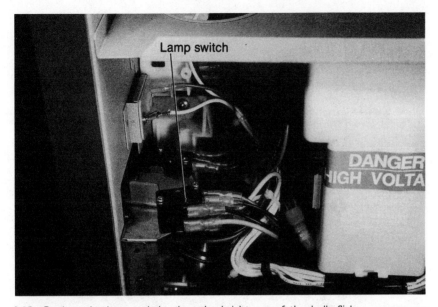

6-13 Replace the lamp switch when the brightness of the bulb flickers or appears erratic. The cavity light switch may be ganged with other interlock switches.

In case the oven lamp flickers with the oven operating, suspect a loose bulb or a defective interlock switch. The light may flicker with the door open and be normal with the door closed. Replace the secondary interlock switch. If the lamp does not go out with the door open or closed, replace the lamp or latch switch. A low flickering lamp with the oven in operation may indicate heavy current being pulled in the magnetron tube circuit.

Don't overlook a defective lamp socket or loose wiring harness when the cavity lamp will not come on. Use the vom and check for 120 Vac across the lamp terminals. The 100 watt pigtail lamp may be clipped across the lamp socket as indicator. Suspect a defective harness or wiring cable if the light appears intermittent when the wiring is moved with a long plastic tool. Lift the cable with a wooden or plastic tool. Do not use a metal tool or screwdriver inside the oven area. Now, open and close the oven door with power connected to the oven.

If the light socket is intermittent or arcs internally, replace it. Unplug the microwave oven and discharge the hv capacitor. Unsolder the leads from the light socket. Check the resistance across the light socket with R×1 scale of the ohmmeter. The meter should read 0 ohms. Replace the socket with actual oven part numbers. To reinstall, reverse the above procedures.

INTERMITTENT OVEN COOKING (CUTOUT)

Suspect a defective thermal protector switch if after the oven has been cooking for several minutes it stops. The fan and turntable motor may be operating except there is no cooking. Although the thermal cutout may be bolted to the magnetron tube, the cutout assembly controls the low-voltage circuits. In some ovens, you may find two separate thermal protector assemblies. One is located on the magnetron and the other on the metal oven cavity (FIG. 6-14).

Oven thermal cut-out

6-14 In some ovens, you may find two separate thermal cutout components. One is mounted on the magnetron and the other on the oven wall. Here, the cutout mounts next to the thermistor in a Quasar oven.

The magnetron thermal protector switch prevents the magnetron from overheating and shuts down the cooking process. Inside the thermal unit, a switch contact connected to a bi-thermal strip opens and closes with heat. If the magnetron tube has an internal leakage or short, the metal shield becomes hot and in turn opens the thermal cutout switch. The oven may start to cook once again after the magnetron has cooled down. Suspect a defective thermal cutout assembly, magnetron, or triac when intermittent cooking is noted.

Check the thermal protector with the vom on the low R×1 ohm scale. Make sure the power plug is pulled and the hv capacitor is discharged. Connect the ohmmeter lead directly across the thermal-switch terminals. You should read a direct short—no resistance. When 2 to 5 ohms resistance is noted, replace the thermal switch. Although the oven may operate, the thermal protector will soon produce erratic heating or cooking (FIG. 6-15).

In case the cooking cycle is cutting on and off within a few seconds, the thermal switch may be checked with a 120 Vac measurement across the terminals. An open thermal unit will indicate the entire power-line

6-15 Here the thermal cutout is located on the magnetron tube body of a Goldstar microwave oven.

voltage (120 Vac). Usually, the thermal switch is in series with one side of the low-voltage circuit. No light or voltage measurement indicates the thermal switch is normal. Be careful with these voltage measurements. Clip the vom leads across the thermal cutout and then fire the oven up (FIG. 6-16). Also, the 100 watt pigtail bulb may be used as a monitor light across the thermal unit.

The thermal protector assembly may be located in a low-voltage control board circuit. Here the thermal switch is in series with the power oven relay. The voltage furnished to the relay by the low-voltage transformer is controlled by the DPC (FIG. 6-17). An open thermal cutout assembly in this type may measure from 20 to 30 volts supplied by the control board.

The thermal protector should be replaced when it is defective or when installing the magnetron tube. Simply remove the two mounting screws and mount the thermal unit on the newly installed magnetron. Replace the connecting wire clips to the thermal component. No harm is done if these two wires are interchanged. The magnetron thermal switch is interchangeable, but not so with the oven thermal assembly.

After 15 minutes you may find the thermal switch will open up in some ovens, indicating the magnetron is overheating, which would normally indicate the cutout or magnetron is defective. First, replace the thermal protector switch (FIG. 6-18). If the new unit opens up after several minutes, check the magnetron for leakage. Both the magnetron and ther-

6-16 Set the VOM or DMM at 300 Vac. Clip the test leads across the thermal cutout terminals. The cutout is defective if any voltage is measured across the two thermal connections.

6-17 The thermal cutout may be located in the low-voltage control circuits. Here, the thermal cutout is in series with the power relay.

6-18   A defective thermal cutout may open after a few minutes of operation, stopping the cooking process. Monitor the power-line voltage (120 Vac) across the two terminals.

mal cutout may be normal. Slip a piece of insulation between the thermal switch and body of the magnetron. Some manufacturers provide thermal insulation kits for this purpose. Replace the magnetron tube if it is running extremely hot.

REPLACING THERMAL PROTECTION

Before testing or replacing thermal protectors, unplug oven and discharge the hv capacitor. Check the thermal protector for open or erratic conditions with the ohmmeter; the meter should read 0 ohms. Remove the wires from the thermal protector. Note how they are connected. Remove the screws holding the defective thermal unit upon the magnetron tube. Install a new thermal protector with the same type. Reverse the above procedures to install it.

OVEN DOES NOT GO INTO COOK CYCLE

In the low-voltage circuits a defective timer switch, thermal protector cook switch, component wiring, or cook relay may cause a no-cook cycle. Suspect an open cook-relay assembly after checking all other components with the ohmmeter. This relay may be called a cook, hold, oven power, or latch relay. The cook relay provides a switch in one side of the power line to the primary winding of the power transformer, completing the circuit. A power relay may operate from the power line or from the low voltage of the controlled assembly. The cook or power relay is energized during the entire cooking cycle.

Often, the cook relay may be heard when energized or you can see the switch point contacts close after the cook button is pushed (FIG. 6-19). No action from the cook relay may indicate voltage is not applied to the relay solenoid winding. Pull the power cord and discharge the hv capacitor. Clip leads to the solenoid terminals and connect to the 300 Vac vom scale. Prepare the oven and push the cook button. No voltage measurement (120 Vac) may indicate a defective interlock micro or cook switch.

6-19 Usually, the cook relay can be heard when energized, or you can see the switch points close. This oven relay is in an early Litton 402 model.

When ac voltage is measured at the solenoid winding and the relay does not energize, suspect an open coil. Discharge the hv capacitor with the power plug removed. Remove one side of the cable connection of the solenoid winding so you will not measure any other resistance in the circuit. No resistance measurement indicates the coil is open. Check for a broken coil wire at the winding. Sometimes these small wires are broken off right at the terminal connections. Replace the oven relay when open or a very high winding resistance is measured. Take an ohmmeter reading across the coil terminals. This resistance may vary between 150 to 300 ohms.

Often oven or auxiliary relays operating from the electronic control board are low-voltage types. You should have a 10 to 24 volt measurement across the solenoid winding. The coil resistance may be from 1k to

2k ohms. Check the manufacturer's service literature for correct resistance. When little or no voltage is measured at the solenoid winding, suspect a defective controller or power transformer.

In very difficult situations, the relay may be energized manually by pushing down on the top section of the relay with a wooden or plastic dowel. Actually, you are making the switch terminals come together so the oven will begin the cook cycle. If the magnetron begins to operate, you know the relay or the applied voltage is at fault. Do not attempt this method in crowded areas. The power relay must be out where you can get to it. Now, if ac voltage is applied to the primary winding of the transformer, you know either the relay is defective or it has no applied voltage.

OVEN KEEPS OPERATING (DEFECTIVE TRIAC)

A defective triac may let the oven operate after the program controller is shut off. In fact, the defective triac may cause many different problems. The gate voltage of the triac assembly is controlled by the controller or DPC. When a voltage is fed from the control board to the gate terminal of the triac, the triac is turned on. In some ovens, this voltage may vary from 4 to 6 Vdc. The triac acts as a switch and connects one side of the power line to the primary winding of the transformer.

The triac may be checked with applied voltage or ohmmeter tests. To determine what component is defective, monitor the ac voltage at the primary of the power transformer (FIG. 6-20). No voltage at the primary winding with the oven in the cooking mode indicates a defective triac, electronic controller, thermal protector, or faulty cable connections. You may assume the electronic control board is working if the unit counts down properly. Quickly check the contacts of the thermal switch with the R×1 ohmmeter scale.

Either the electronic controller is not feeding voltage to the gate terminal or the triac is defective. To check the triac in the circuit, measure ac voltage across one side going to the power transformer and the other to the power line (FIG. 6-21). When 120 Vac is measured across these two terminals, the voltage is present but the triac is not switching. A 100 watt pigtail lamp may be used as an indicator across these triac terminals.

To determine if the hv circuits are operating, simply clip a lead across these triac terminals (1 and 2). Always pull the power plug and discharge the hv capacitor before reaching into the oven. Use clip leads when connecting leads or test instruments to the oven circuits. Now, with the clip lead in place, start the oven up. The oven should produce heat and cook without any time control (FIG. 6-22).

Remove the suspected triac assembly. You may find several different triac units or modules in different microwave ovens. Usually, the triac assembly is bolted to the oven cavity wall. All terminal leads must be removed before checking the triac with the ohmmeter. Mark down each wire connection. You will find a different color code of each connecting

6-20 To determine what component is defective, monitor at the primary winding of the power transformer (120 Vac). No voltage indicates problems with the triac, electronic controller, thermal protector, oven relay, or cable connections.

6-21 To check the triac in the circuit, measure the ac voltage across one side going to the power transformer and the other to the power-line. If there is 120 Vac across these terminals, suspect a defective triac or controller gate voltage.

cable. Interchanging the triac leads may damage the new triac or control board. Often the gate terminal is the smallest space-type lug and cannot be interchanged with the others.

The triac module or assembly connections are easy to trace out. The MT2 terminal connects to one side of the power transformer (FIG. 6-23).

6-22 Here is a shorted triac that will not turn off the oven with the control unit.

6-23 The MT2 terminal of the triac connects to one side of the power transformer. Terminal MT1 connects to one side of the power line in a Norelco microwave oven unit.

Terminal MT1 connects to one side of the power line. The gate terminal (G) connects to the connecting wire from the electronic controller or control board. The gate voltage supplied from the controller in the Norelco R7700 oven is a −5 volts.

Check the continuity between any triac terminals with a vom, vtvm, digital vom, or the diode test of a digital meter. Switch the meter to the diode test of the digital meter and read the resistance between G and MT1. This reading will be somewhere around (.554Ω) in either direction (FIG. 6-24). Reverse the test leads and the reading should be the same for a normal triac. You will not read any resistance between other combinations of two terminals for a normal triac. If a reading other than gate to MT1 is noted, the triac is leaky.

6-24 The digital relay of a DMM diode tester should be somewhere around 0.554 ohms in either direction at terminals G and MT1. No resistance should be measured between any other terminals with the diode test of a normal triac. Usually, the shorted triac is leaky between MT1 and MT2.

The low-ohms scale of a vtvm will measure around 100 ohms between gate and MT1. In a Hardwick oven, the same terminal resistance may be 22 ohms. A reading above 10 megohms from MT1 and MT2 is normal. Replace the triac assembly when a 5 megohm or lower measurement is noted between any other terminals than MT1 and G. A digital ohmmeter test between MT1 and G may read 100 to 110 ohms. It's best to check each triac reading with the manufacturer's data, if handy. Usually, the defective triac will appear shorted and leaky. Very seldom does the triac go open.

When the oven turns itself off, the fuse is blown and will not start up again until the fuse is replaced. Sometimes the oven may run 10 to 15 minutes before this condition occurs. You may waste a lot of service time on this one. Check the oven and circuit for a defective triac assembly and replace it.

A defective triac may let the oven hum under load with the oven turned on. You just hear a humming noise and have no cooking. You may notice the electronic control board is counting down rather fast. The triac does not turn on the voltage to the primary winding of the power transformer. Simply replace the leaky triac with the original type.

You may find another type of triac assembly subbed or modified from the original one. No doubt the manufacturer has found the new triacs are reliable and should be installed when the old one is found defective. The main thing to remember is to connect each wire as shown in the new schematic. Improper connections of the triac may damage the control circuit board. With triacs marked MT2, MT1, and G1, MT2 goes to the power transformer, MT1 to one side of the power line, and G goes to the gate connection from the control board.

REVERSED TRIAC CONNECTIONS

Before replacing a suspected triac, mark down the color lead wire connections. In the Norelco ovens, the MT2 connection has a blue wire, the

MT1 has a white wire, and the gate is silver. With the Hardwick ovens, the triac MT1 lead is a white wire, the MT2 is a black wire, and the gate terminal is a red wire. Remember, the color code of the various triac lead wires may be different in various ovens. If MT2 and MT1 wires are reversed, the microwave oven will operate 180° out of phase and may damage the control microprocessor assembly circuits. Always keep the gate wire connected to the triac. If the gate wire is removed or accidentally knocked off, the microprocessor may be damaged in the control circuits.

LAMP TRIAC TESTS

You may check the oven triac assembly with a lamp-type tester (FIG. 6-25). Build the tester with rubber-coated alligator clips so you do not get shocked. Clip the leads to the triac with the power plug pulled on the tester. The triac leads must be removed for these tests. It's best to remove the triac and test on the bench.

6-25 Make you own ac triac tester out of a few ac components.

After the triac is connected, check it for a short between MT1 and MT2. Just plug the tester in and if the lamp does not light up, the triac is not shorted. Now, press down on the momentary SW1 and rotate R1 until the bulb lights. Let up on the SW1 and if the light stays on, the triac is normal. If the light does not light up at all, suspect an open triac assembly.

EVERYTHING LIGHTS UP BUT NO COOKING (MEAT PROBE)

In some ovens, the temperature probe must be inserted when the cooking switch is turned to the meat probe operation. If the meat probe is not in use and the switch remains in that position, the next time the oven is turned on it will not cook. Undoubtedly, this situation is caused when the probe is used and not switched to the probe position (this may occur in a few ovens). The oven may come in for repair with nothing wrong—only turn off the probe switch.

Improper meat probe temperature cooking may be caused by a defective temperature probe, probe jack, or faulty wiring. The temperature probe may be tested with a 100 kilohm scale. Connect the ohmmeter leads to the temperature probe male jack. This jack looks like any standard earphone jack with the point and common terminal connected to the meter (FIG. 6-26). Normal resistance may be measured from 30 to 75 kilohms. If infinite or extremely low resistance is measured, replace the temperature probe.

6-26   The temperature probe jack looks like the earphone jack with the common and point terminals connected to the meter. Normal resistance may be from 30k to 75k ohms.

Check the temperature probe once again with the ohmmeter leads connected and immerse the tip of the probe in hot water. In case the ohmmeter resistance decreases with the probe immersed in hot water, check the probe jack for problems. However, if the resistance does not decrease, replace the temperature probe. Never let the probe lay on the metal oven floor while the oven is in operation. Remove the probe when not in use.

A defective probe jack may prevent the oven from operating. You may find some probe jacks have a short circuiting terminal while others are open (FIG. 6-27). In a Norelco oven there is a short-circuit jack. Simply use the R×1 scale of the vom and make continuity tests of the terminal connection and cable wires. A typical probe test chart is shown in FIG. 6-28.

DIFFERENT TEMPERATURE PROBE TESTS

Different microwave ovens may have probes that have different elements resistance. The defective probe may appear leaky or open. Measure the resistance across the probe jack. If the resistance is lower than normal, the probe element is leaky. Check the wiring and plug for open measurement. The various oven probe resistances at room temperature are shown in TABLE 6-1.

Panasonic's probe thermistor tests   To test the sensor probe's thermis-

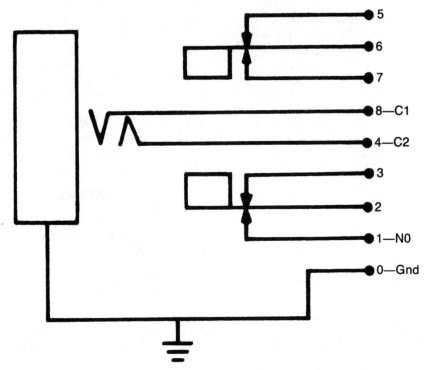

6-27 A typical probe socket with the short-circuitry terminals connections.

Condition of probe	Terminal Test Points	Normal Indication
Probe Out	G to C1	Open
	C2 to N0	Open
Probe In	G to C1	30 k to 75 k ohms
	C2 to N0	Continuity

6-28 A typical probe test chart is shown for the various connections.

tor, fill a container partially with lukewarm water and place a thermometer in it. Then take the sensor probe out of the oven and place it in the container next to the thermometer so they both will sense the same temperature.

Now take another container with boiling water and gradually pour this into the first container to achieve graduated temperature steps. At

Table 6-1 The different oven temperature probe's resistance at room temperature.

Manufacturer	*Temperature Probe Resistance in Ohms*
Amana	50,000
Hardwick	50,000
Norelco	55,000
Samsung	46,600
Sharp	49,500

Table 6-2 Panasonic sensor thermistor probe test with different temperatures and resistance.

Temp.		Thermistor Resistance Nominal (kΩ)	Temp.		Thermistor Resistance Nominal (kΩ)
(°C)	(°F)		(°C)	(°F)	
30	86	44.4	60	140	13.7
35	95	35.9	65	149	11.5
40	104	29.3	70	158	9.7
45	113	24.0	75	167	8.2
50	122	19.8	80	176	7.0
55	131	16.4	85	185	6.0
			90	194	5.1

each temperature step, you should read the corresponding resistance with the ohmmeter (TABLE 6-2). If not, the probe is defective and should be replaced.

Check to see if the probe thermistor is working under cooking conditions. Set the oven cooking temperature at 150°F. Fill the testing cup with about 9 ounces of tap water. Place the container and probe in the oven. Time the oven to where it shuts down at 150°F. Now check the temperature of the water with the thermometer, the water temperature should be close to 150°F. Check the resistance of the probe to see if resistance has changed. If not, replace the temperature probe.

DEFECTIVE BROWNING UNIT

A convection oven has two types of cooking methods enclosed in one oven. Often, the ac coil heating element is located at the top of the oven and in some cases may be lowered over the food to be cooked. When the element is enclosed, a rotating fan blows the hot air on the food. You may

6-29   You may locate a browning or cooking element at the top of the oven's cavity. In this Sharp model a fan blows the heated air to the oven cavity.

find a browning or cooking element at the very top of the oven cavity (FIG. 6-29).

When the heating element does not appear to become hot, suspect a burned cable, bad switch, poor wire to element connection, or an open heating element. The heating element may be checked with the R×1 scale of the vom. This element is always insulated from the metal oven cavity. No ohmmeter reading should be obtained from the heating element to the cavity walls.

Check the screw terminals where the insulated cable connects to the end of the heating element. Poor terminal connections or a burned off wire prevents the element from heating up. Often, the screw connections will become burned and brown-like, indicating poor terminal connections. If the cable is burned off and is too short to make adequate connections, replace with the original cable harness from the manufacturer—these are a special type asbestos insulated cables. Loose or sharp points on element mounting screws may cause arcing inside the oven when the microwave oven is cooking. Also, check for sharp edges or loose screw terminals in the oven cavity (FIG. 6-30).

In a Sharp convection Model R8310 oven, a round heating element is located at the top center of the oven cavity. A fan motor drives a long belt to the center of the oven, rotating an enclosed fan blade. The fan pushes the hot air around convection passages to cook the food. A thermistor is mounted inside for correct temperature of the control unit (FIG. 6-31). The heating element is controlled by a triac assembly within the control unit (FIG. 6-32). When the convection timer is set, a damper control lever assembly must shut off the exhaust part completely. With the convection timer in the off position, the damper must be opened fully for magnetron operation.

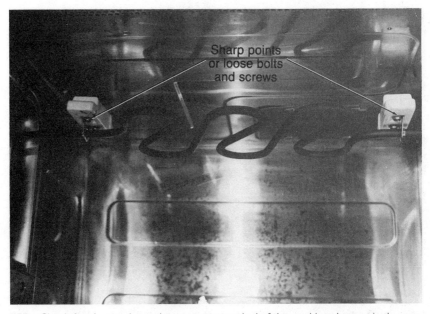

6-30 Check for sharp edge or loose screw terminal of the cooking element in the oven cavity. These connections may cause arcing or fire-over points when cooking with the microwave oven. Today, anodized metal shelves in the oven do not show signs of arcing.

6-31 A thermistor is mounted for correct temperature of the control unit in a Sharp R8310 model. The thermistor controls the triac assembly which in turn shuts down the magnetron tube.

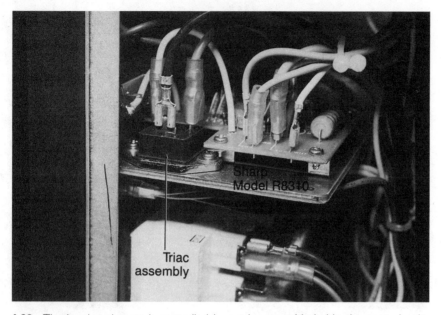

Triac
assembly

6-32 The heating element is controlled by a triac assembly inside the control unit.
When defective, replace the entire control assembly.

While in the convection cooking mode, the convection timer
switches one side of the power line from the primary winding of the
power transformer to the temperature control unit. The 120 Vac is supplied to the heating element through the triac of the temperature control
assembly. When the temperature of the negative-coefficient type thermistor rises above the selected temperature, the circuit to the heating element
is cut off by the control unit. As the temperature drops below the selected
temperature, the 120 Vac is supplied to the heating element.

In case the convection section of the oven cavity is not heating,
check for 120 Vac across the heating element terminals. The heating element is open when power-line voltage is measured at the terminals. Pull
the power plug and measure for 9.1 to 9.6 ohms resistance. No continuity
between the heating element terminals indicates that the element is open.
Higher resistance than normal may cause a poor or slow cooking process.
Check the resistance of the heating element to chassis ground. This resistance measurement should be above 600 kilohms. Replace the heating
element if a lower reading is found.

Suspect a defective temperature control unit if no ac voltage is found
at the heating element terminals. Check for loose or poor wire connections of the control assembly. Disconnect the leads from the thermistor.
The resistance between the temperature thermistor terminals should be
somewhere between 140 to 390 kilohms. Check for 120 Vac at the input
and output terminals of the temperature control unit. Replace the entire
temperature control unit when 120 Vac is applied and there is no output
voltage to the heating element.

Panasonic heating element replacement   In models NN-9807 and NN-9507, only one heating element is located in the top of the oven. While in the NN-8507/NN-8807/NN-8907 CPH oven, a lower and upper heating element is powered by relays controlled by the DPCs. After finding either element open or shorted, the unit must be replaced with correct part number.

In the NN-9807 and NN-9507 oven, replace the heater as follows:

1. Remove the lead wires from the lead wire holders (FIG. 6-33).

Pulley of
circulation fan   Heater unit

Lead wire   Oven temp   Belt   Heater
holders   sensor   fan motor

6-33   Remove lead wire, holders and belt from pulley of circulation fan of Panasonic NN-9807 and NN-9507 oven to remove the top heating element (courtesy of Matsushita Electric Corporation of America).

2. Remove the belt from the pulleys of the circulation fan.

3. Remove the two screws holding the temperature sensor.

4. Remove the two lead wires from the heater terminals by removing the screws (FIG. 6-34).

5. Remove the six screws holding both sides of the heater unit and lift the heater unit carefully.

6. Remove the two spring nuts from the heater terminals.

7. Remove the two screws holding the heater supports and detach the heater (FIG. 6-35).

CONCLUSION

Measure the ac voltage (120 V) across the primary winding of the power transformer to determine if problems are within the low-voltage or hv circuits. No ac voltage here indicates low-voltage problems. Leave the vom connected as the voltage monitoring device. Use lights and various

Spring nuts

Heater terminals

6-34 Remove the two lead wires from the heater terminals in Panasonic ovens to remove the top convection heater (courtesy of Matsushita Electric Corporation of America).

Heater supports

Heater   **Circulation fan**

6-35 Remove the two screws holding the heater supports in order to detach the heater in the NN-9807 and NN-9507 Panasonic ovens (courtesy of Matsushita Electric Corporation of America).

motors to determine what part of the low-voltage circuit is functioning. The defective component may be between the last working component and the ac voltmeter. When the power-line voltage is measured at the primary winding of the power transformer, with no heat or cooking, suspect a defective component in the hv circuits.

Chapter 7

Troubleshooting high-voltage problems

The high-voltage (hv) basic circuits consist of a magnetron, hv capacitor, diode, and power transformer. Power-line voltage is applied to the primary winding of the power transformer and then boosted with a voltage-doubler circuit. When a high dc voltage is fed to the magnetron, the tube oscillates sending rf energy from the antenna to the waveguide assembly. The rf energy enters the oven cavity and cooks the food in the oven.

THE MAGNETRON

A magnetron tube consists of a heater, cathode, metal cylinder block, and antenna (FIG. 7-1). The inner cylinder represents the heater and cathode. The heater may be called the filament in some microwave ovens. An outer cylinder represents the anode block, strap ring, and vanes. When the heater voltage is applied, the cathode emits electrons. With a higher voltage on the anode block area, electrons start to travel from the cathode to the outer cylinder.

The magnetic field produced by an external magnet makes the electrons spin faster about the inner cylinder. These spinning electrons between cathode and anode generate the microwave power at 2,450 MHz. A small antenna picks off the rf energy that is transmitted to the oven cavity via the waveguide assembly.

In FIG. 7-2 the various parts of the magnetron tube are shown. The

7-1 The magnetron tube consists of a heater, a cathode, a metal anode, and an antenna assembly. Metal fins or vanes provide cooling of the magnetron.

7-2 Here is a breakdown of the components of the magnetron. The long heater leads enter the bottom sealed areas. Two strong magnets provide a highly concentrated magnetic field around the metal anode.

large heater terminals feed into the bottom of the magnetron. The metal fins are spot welded to the outer anode cylinder which operates at ground potential. Besides heater voltage, a high negative dc voltage is fed to the heater or cathode terminals. Notice the small metal antenna enclosed within the glassed area. The air within the magnetron is pumped out, like any vacuum tube.

A heavy magnet is placed around the anode cylinder to provide a highly concentrated magnetic field. Here, two smaller fixed magnets are used on each end while powdered iron core magnets are found in the new magnetron tubes. Be careful when replacing or working around the magnetron tube. These strong magnets may pull a metal tool from your hand and in the process break the glass area. Magnetron replacement is one of the most costly microwave oven repairs.

To operate, the magnetron must have a very high dc voltage (up to 4,500 Vdc) and a heater or filament voltage of 3.1 volts ac (Vac). When two separate power transformers are found in the older ovens, the smaller one may furnish only ac voltage to the heater circuit, while the larger power transformer completes the voltage doubler circuit. Usually, the newer ovens have only one power transformer furnishing ac to the voltage doubler circuit with a separate heater winding.

THE VOLTAGE-DOUBLER CIRCUIT

Ac voltage is supplied to the voltage doubling circuit by the secondary winding of the large power transformer. The power transformer may sup-

7-3 Extreme care must be exercised while working around the high-voltage areas. Always discharge the hv capacitor before taking ohmmeter tests or testing any component in the oven.

When ac voltage is supplied to the voltage doubler circuit, the hv capacitor is charged. As the ac current reverses the direction of current flow, the hv diode prevents the capacitor from being discharged back through the diode. The peak voltage may be from 2,000 to 6,000 volts.

ply 1,800 to 3,000 ac volts to the voltage doubler circuit. Of course, the output dc voltage is not double the ac input voltage due to the loss in the circuit (FIG. 7-3).

The hv diode rectifies the ac voltage and provides a high dc source to operate the magnetron tube (1,800 to 4,500 Vdc).

The high voltage developed within the voltage-doubling circuits is very dangerous and cannot be measured with ordinary test equipment. Special test equipment is needed. Never attempt to measure the high dc voltage with a pocket vom. Use either a special hv dc meter, vtvm with hv probe or a specially designed microwave oven Magnameter. Notice the danger hv warnings around the magnetron tube area (FIG. 7-4). Always discharge the hv capacitor before attempting to clip the test leads to any hv component.

7-4 Notice the high voltage danger sign around the magnetron tube area. Most manufacturers list the danger areas.

Panasonic's hv circuits   In most of the microwave ovens, the high voltage is generated by the action of the diode and the charging of the hv capacitor. This circuit is called a *half cycle double circuit*. This circuit is commonly used because it is economical as a smaller transformer and a capacitor can be used to produce the high voltage.

The typical half cycle doubler circuit with the capacitor and diode

HIGH VOLTAGE CIRCUIT

7-5 A typical half cycle doubler circuit with capacitor and diode in the Panasonic ovens (courtesy of Matsushita Electric Corporation of America).

are connected in the hv transformer secondary (FIG. 7-5). Generated from the filament winding on the hv transformer, 3.3 Vac is applied to the magnetron filament through noise suppression choker and capacitors. Two chokes and capacitors, enclosed within the magnetron shield case, prevent microwaves from effecting radios and television sets.

The ac voltage of approximately 2,000 volts or more, depending on the output power of the microwave oven, is generated from the secondary winding of the hv transformer. The capacitor charges through the diode during the first positive cycle of the ac from the transformer (FIG. 7-6). The charge path of the capacitor is shown by the dash lines. During

7-6 The capacitor charges through the diode during the first positive cycle. Follow the dotted lines and arrows (courtesy of Matsushita Electric Corporation of America).

the capacitor charging time, the magnetron is off since the diode shunts it. The capacitor charges to approximately 2,000 volts or more.

During the negative half cycle, the voltage on the capacitor and the voltage across the transformer secondary winding are combined and applied across the magnetron anode, shown by the solid lines (FIG. 7-7). This resultant potential of approximately 4,000 volts is used to oscillate the microwave from the magnetron. Notice that the magnetron is pulsed on and off at a rate of 60/50 cycles per second depending upon the line rate used.

7-7 During the negative half cycle, the voltage on the capacitor and the voltage on the transformer secondary winding are applied across the magnetron anode, shown in the solid lines (courtesy of Matsushita Electric Corporation of America).

Some models have a varistor across the diode. This varistor protects the circuit in case of hv surges generated at the magnetron. The varistor normally has a very high resistance. However, when a high pulse (approximately 10kV) is applied across it, the internal resistance decreases. This low resistance characteristic is used to shunt the magnetron and diode.

Some models do not have this varistor externally. These models have a diode called an *avalanche type*, which is similar to the varistor but built into the molded diode assembly.

In order to discharge the hv capacitor when the oven is turned off, a bleeder resistor (9 or 10 megohm) is provided. The time required for complete discharge is approximately 30 seconds. Be careful: Do not assume that the bleeder would have discharged the capacitor. Always discharge the capacitor before working on the hv circuit or magnetron. In some ovens, two hv capacitors are used for high voltage output. The power selector switch must be set in the high power position to discharge both capacitors.

THE HIGH-POWER TRANSFORMER

The high-power transformer may have a primary and two secondary windings. One of these secondary windings supplies ac voltage to the heater or filament circuit of the magnetron. The large hv winding supplies high ac voltage to the voltage-doubler circuit (FIG. 7-8). You may find a separate heater transformer in some ovens. The defective power transformer may go open, have shorted turns within the hv winding, or arc through the hv winding to the metal core area. Actually, when no high dc voltage is present at the magnetron, the power transformer may be checked with the ohmmeter. Before taking resistance measurements, pull the power cord and discharge the hv capacitor. The primary winding may have a resistance of 0 to 2 ohms depending on the oven transformer and what ohmmeter you are using. Most primary windings have a resistance under 1 ohm. A resistance measurement with a digital vom may read a fraction of 1 ohm.

7-8 The large power transformer supplies high ac voltage to the voltage-doubler circuits. Another low ac voltage winding is provided for the heater or filament circuit of the magnetron.

The secondary hv winding may measure from 50 to 100 ohms. Some manufacturers give a reading of 20, 50, and 70 ohms for the secondary winding resistance. Use the R×1 scale of the vom (FIG. 7-9). The filament

7-9 The secondary hv winding may measure a resistance from 50 to 100 ohms. Use the R×1 ohm scale of the VOM.

or heater resistance may show a short or 0 ohms. Always pull off one of the heater leads to the magnetron for correct resistance measurement. If not, the heater of the magnetron and power transformer winding is in parallel and you may have an open transformer winding or connection, resulting in erroneous filament or heater measurement (TABLE 7-1).

In some older ovens, you may look down inside the top of the magnetron and see the heater light up, while most are covered with a metal guide cover and will not show the heater light. To determine if the heater of the magnetron or heater winding of the transformer are open and have a higher than normal resistance reading, take an accurate resistance measurement. If the heater winding remains in doubt, disconnect the heater leads from the magnetron and take an ac voltage reading. When checking the voltage, keep meter leads and hands away from the hv circuits. Set the voltmeter on a book or service manual away from the oven. Clip the meter into the circuit with test leads. The heater voltage of most ovens ranges from 2.8 to 3.6 Vac (FIG. 7-10).

Panasonic's hv transformer tests   Remove the connections from the transformer and check the winding continuity. The spade lugs pull off the

Table 7-1   Resistance measurements of power transformer in the various ovens in ohms.

Manufacturer	Primary	Secondary	Filament
Amana	0–1	109–130	0–1
Hardwick	0.16	60	0–1
Panasonic	0–1	80–120	0
Samsung	0	50–70	0
Sharp	0–1	53	0–1
Whirlpool	0.2	55–61	0

7-10 The heater or filament voltage of most ovens ranges from 2.8 to 3.6 Vac. Disconnect the heater winding terminals from the circuit for this measurement. Clip the low ac voltmeter directly to these two leads.

primary and secondary winding terminals. The secondary hv winding should have a cold resistance measurement of approximately 80 to 120 ohms. The filament winding should have a measurement of approximately 0 to 1 ohm (FIG. 7-11). Before making any resistance or voltage measurements within the hv section, discharge the hv capacitor.

7-11 Checking the power transformer resistance measurement in the Panasonic NN-9807 hv section (courtesy of Matsushita Electric Corporation of America).

FILAMENT TRANSFORMER TESTS

Some of the microwave ovens have a separate filament transformer in the hv circuits. In other words, there are two separate transformers providing voltage for the magnetron (FIG. 7-12). The primary winding should have a resistance from 0 to 20 ohms, and the filament terminals should have a

7-12 Some of the microwave ovens have a separate filament power transformer.

resistance of less than 1 ohm. This filament transformer has hv insulation just like the large power transformer, as it is connected to the hv dc source. Of course, it is physically much smaller compared to the hv transformer.

Microwave ovens with digital controls have a low-voltage transformer to supply ac voltage to the control circuits. You may find several secondary windings furnishing 2.5 to 3.3 Vac and 20 to 28.5 Vac. The low-voltage power transformer may be checked with ac voltage and resistance measurements. Check the manufacturer's oven schematic for correct voltage and resistance readings.

Besides open conditions, the power transformer may cause intermittent and erratic oven operations. Check the transformer filament lug terminals on the magnetron heater connections for burned crimped connections and wiring. When found, replace the crimped connections if possible. Sometimes these connections may be cleaned up and soldered to the component terminals.

Poorly crimped solder lugs on the power transformer enameled wiring may cause intermittent or dead oven conditions. Often, these lugs are crimped right through the enameled wires and in time make a poor connection. Simply scrape the wire with a pocket knife where the crimped connection ties to the wire winding (FIG. 7-13). Apply rosin core soldering paste to the scraped area. Now, solder the crimped lug to the scraped wire. Always check these transformer connections when you find an oven operates erratically or intermittently.

Scrape back then tin
and solder

To power
transformer

Enameled
wire of
transformer

Spade crimped
connection

7-13 Intermittent oven operation may be caused by a poorly crimped enameled wire connection. Scrape the enamel off the copper wire, then tin and solder it to the crimped connector.

Panasonic's positive lock connector The positive lock connector is a specially designed locking connector. You will find this connector in many lead wire connections. To remove the connector, pull the lead wire by pressing an extruded lever in the center of receptacle terminal (FIG. 7-14).

How to disconnect
positive lock connector

(1) Push

(2) Pull

Locking
clip

Positive lock connector

7-14 Panasonic positive lock connections found on the magnetron and the thermal cutout (courtesy of Matsushita Electric Corporation of America).

In the NN-9807 oven, when connecting two filament leads to the magnetron, be sure to connect the lead wires in the correct position. The lead wire of the hv transformer should be connected to the F terminal, and the lead wire from the hv capacitor should be connected to the FA terminal (FIG. 7-15).

No dc voltage applied to the magnetron may be caused by a shorted hv winding of the power transformer. First, discharge the hv capacitor. Now, measure the resistance between the heater terminal on the magne-

Thermal cutout

FA   F

Air guide A   **Magnetron**

7-15   Correct filament connections on the Panasonic ovens (courtesy of Matsushita Electric Corporation of America).

tron and chassis ground. One side of the power transformer secondary winding is connected to the oven chassis ground, while the other end connects to the heater circuit. You may find in some ovens a *high* and *low* connection providing a method to increase or decrease the amount of power produced by the magnetron. Normal resistance should be measured from the high side of the secondary winding to chassis ground. In case a very low measurement is found, disconnect the hv transformer lead from the magnetron heater terminal. Now, measure the transformer's hv winding. Compare the reading to those given in the manufacturer's literature. A lower than normal reading indicates shorted turns in the transformer or the winding has arced over to the metal core area. If the transformer winding is normal, suspect a shorted or leaky magnetron tube.

You may find the power transformer has a loud buzz when the oven is operating. A low hum noise is normal. Although the oven operates correctly, the loud buzzing noise may be very disturbing to the operator. You may find the transformer buzz is intermittent only when the oven is under a cooking load. Replace the noisy transformer with one having the original part number. Since these transformers are dipped after assembly, it's impossible to tighten them up to stop the noisy condition.

ARCING HV TRANSFORMERS

You might hear a few loud cracking noises before the fuse opens up, or you might smell something warm. The hv transformer may arc over between windings or from the high voltage side to the metal core. Take a quick continuity resistance measurement between one side of the hv capacitor and ground. The symptoms may be similar to those of a shorted hv capacitor or diode. Replace the power transformer with the exact part number.

THE HV CAPACITOR

When the hv capacitor goes open, shorted or leaky, no high voltage will be available at the magnetron. Before checking the capacitor or any component in the oven, discharge the hv capacitor. Be safe, always discharge the capacitor. Use a couple of insulated screwdrivers for this purpose.

The hv capacitor is mounted close to the power transformer and magnetron circuits (FIG. 7-16). These oil filled capacitors have a capacity rating from 0.64 to 1μF with a 2 to 3 kilohm voltage rating. In early models, the physical size of these capacitors was quite large compared with those used today (FIG. 7-17). Many of these capacitors may be substituted from other microwave ovens.

7-16   The hv capacitor is mounted close to the power transformer and magnetron circuits. These oil-filled capacitors have a capacity rating from 0.64 to 1.0 UF.

In some circuits, you may find a bleeder resistor across the capacitor terminals. This resistor bleeds off the capacitor charges after the oven is turned off (FIG. 7-18). Before checking the capacitor for shorts or leakage, disconnect one side of the 10 meg resistor. Set the ohmmeter to the 10k scale and preferably use a meter with a 6 volt battery.

Remove connecting cables to one side of the capacitor and place the meter leads across the terminals. The capacitor will charge up with the meter hand going up scale and then slowly discharge. Reverse the meter

7-17 In the early ovens, the hv capacitors were very large in size. The one on the left is from a new oven hv circuit.

7-18 A large 9-10 megohm resistor may be found across the capacitor terminals to discharge the capacitor after the oven is shut off. Don't take a chance: discharge each hv capacitor before attempting to repair it.

leads and the same process should occur. Replace the hv capacitor if it will not charge.

You may find an hv capacitor with only a few ohms of resistance between the connecting terminals. Discard the capacitor because of leakage or shorted conditions. Test the capacitor for leakage between the terminals and the outside metal case. No continuity measurement should be shown between can and terminals. Any reading less than 100k indicates a faulty capacitor. If in doubt, sub another oil filled capacitor. Here are a few actual capacitor failure problems:

Dead—No cooking

The rest of the oven operated perfectly except no cooking or heat. No dc high voltage was measured at the magnetron. When discharging the capacitor for tests, no arcing or snapping noise was heard. At first, the hv diode was checked and appeared normal. One terminal connection was removed from the capacitor with the ohmmeter clipped on. The capacitor would not charge in either direction. Another hv capacitor was clipped into the circuit and the oven came on. The open capacitor was replaced with the manufacturer's exact replacement part.

No cooking—No heat

Again all functions of the oven except the water test appeared normal. The water temperature was cold. A dc high voltage measurement of the magnetron indicated no high voltage, although 120 Vac was measured at the primary winding of the power transformer. Since the hv diode caused more trouble in the voltage-doubling circuit, it was checked right away. The diode was normal. A resistance measurement between the filament of the magnetron and ground measured 72 ohms. Either the magnetron or capacitor was leaky. The capacitor measured 1.3 ohms across the terminals and was replaced (FIG. 7-19).

7-19 Check the hv capacitor for leaky conditions. In this circuit, the defective capacitor had only 1.3 ohms across the terminals.

Popped and went out

The customer complained she was cooking with the oven and it gave out a loud pop and then quit. A 15 amp fuse was replaced, but still no heat or cooking. Instead a measuring a high dc voltage at the magnetron, low voltage was present. The hv diode appeared normal. The leads from one side of the capacitor were removed. Another hv capacitor was clipped into the circuit and the oven began to cook. A resistance measurement across the capacitor terminal indicated the capacitor had internally broken down. It's best to remove the hv diode from the circuit when making high-resistance measurements.

Panasonic's hv capacitor tests Before attempting to take resistance or voltage measurements in the oven, discharge the hv capacitor. An extremely low crack may indicate the capacitor or diode is open. Clip the meter across the hv capacitor terminals.

1. Check the continuity of capacitor with the meter on its highest ohm scale.
2. A normal capacitor will show continuity for a short time and then indicate 9 megohms once the capacitor is charged.
3. A shorted capacitor will show continuous continuity.
4. An open capacitor will show a constant 9 megohms.
5. Resistance between each terminal and the chassis should be infinite.

9 VOLT METER CAPACITOR TEST

Several oven manufacturers recommend checking the hv capacitor with a meter having a 9 volt battery. Discharge the hv capacitor and remove the leads to the capacitor terminals. Set the ohmmeter on R×10,000 scale. The meter hand should momentarily deflect upward to show continuity, then return to show infinite ohms. Now, reverse the leads. The meter should give the same indications. The same should hold true with a DMM (FIG. 7-20). The numbers will go quite high and slowly discharge downward.

If the meter hand goes to a low ohm resistance, the capacitor is leaky. With an infinite measurement and no hv arc-over, the capacitor may be open. If all readings appear normal on the capacitor, check the hv diode, magnetron, or hv power transformer.

THE HV DIODE

The silicon hv diode and magnetron are the most troublesome components in the hv circuit. These diodes come in many sizes and shapes. They may be interchanged from other ovens if they can be mounted properly. Often, the hv diode is mounted quite close to the hv capacitor (FIG. 7-21).

7-20 Checking the hv capacitor with a DMM. Discharge the hv capacitor before connecting any instrument to it.

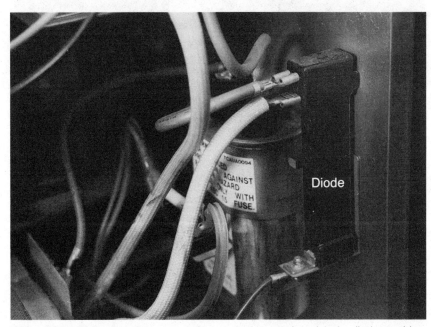

7-21 Often the hv diode is mounted close to the hv capacitor. A leaky diode may blow the 15 amp fuse.

A defective diode may keep blowing the 15 amp fuse or produce a no heat, no cook condition. Before attempting to check the diode, discharge the hv capacitor. If the body of the diode appears quite warm, replace it (FIG. 7-22). In most cases, the hv diode shorts or becomes leaky.

7-22   Replace the capacitor when the body appears very warm. Do not touch the diode or any component in the oven until the capacitor is discharged. The finger points at the defective hv diode.

The suspected diode may be checked with the R×10k ohmmeter range. Preferably use an ohmmeter with a 6 or 9 volt battery. Do not use an ordinary vom or infinite resistance may be read in both directions. Isolate the diode by disconnecting one lead from the circuit. Clip the test leads across the suspected diode. The diode may read several hundred ohms. Now, reverse the test leads. No reading should be obtained (infinity).

Replace the diode when a low-resistance measurement or the meter shows continuity in both directions. If the diode measures below 150 kilohms, replace it. A vtvm or digital vom are ideal to check for a leaky silicon rectifier. If in doubt, clip another diode in the circuit. Watch for correct polarity. The positive or cathode terminal is always at ground potential. When the hv diode cannot be found in older models, check the magnetron and then it turns out to be a defective diode inside the magnetron shield (FIG. 7-23).

Norelco hv diode test   A meter with less than a 6 or 9 volt battery is usually not adequate for checking front-to-back resistance of the diode. The meter should be checked with a diode known to be good before judging a

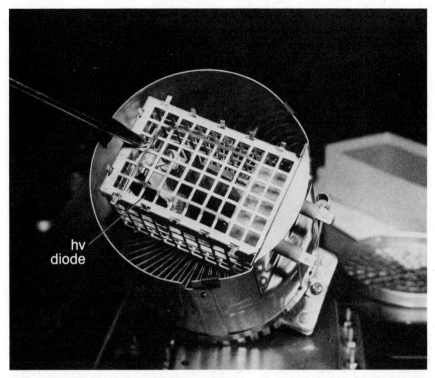

hv
diode

7-23 When the hv diode cannot be found in older ovens, look down inside the magnetron cage. Here, the diode is located inside the shielded cage of an early Litton
microwave oven.

diode to be defective because of an infinite resistance reading in both
directions.

Disconnect the two leads of the diode. With the ohmmeter set on the
highest resistance scale, measure resistance across two diode terminals
(FIG. 7-24). Reverse the meter leads and again observe the resistance readings. A normal diode will indicate infinite resistance in one direction; a
deflection of the meter needle towards the 0 marking is indicated in the
opposite direction. If continuity is indicated in both directions or if an
infinite resistance is read in both directions, the diode is probably defective and should be replaced. Suspect a leaky diode when the body
appears warm.

HOW TO CHECK THE HV CIRCUITS

Only two test measurements are needed to check out the hv circuits. Clip
the ac voltmeter across the primary winding of the power transformer (1
and 2). When 120 Vac is monitored at these connections, you know all
the low-voltage circuits are functioning (FIG. 7-25). If any other problems
exist, the hv circuit and magnetron must be defective.

For the last test, measure the dc high voltage from the filament or

7-24 Measuring the resistance across the hv diode with the diode leads removed from the oven circuits with a DMM.

heater connections to chassis ground. Since the heater terminal connections are covered with plastic insulation, the hv test can be made right across the hv diode. Use either an hv probe or meter such as the Magnameter. Usually proper high voltage across the hv diode indicates the hv circuits are functioning, except that the magnetron may be defective.

7-25 When 120 Vac is found on Terminals 1 and 2 of the power transformer, you know that the low voltage circuits are okay. A high dc voltage measured across the heater terminals to ground indicates the voltage-doubler circuits are functioning.

When the hv circuit and magnetron are working correctly, you will hear a loud hum from the power transformer. Sometimes, the oven lights may dim a little when the magnetron is drawing current. No transformer hum may indicate a defective magnetron or no high voltage. Pull the power plug and discharge the hv capacitor.

In case a dc hv meter is not available or you are afraid to take hv measurements, the hv circuits may be checked in the following manner (some oven manufacturers say it is not necessary to make hv measurements in the microwave oven circuits).

Measure the resistance of the hv transformer winding (50 to 100 ohms). If this measurement is normal, go to the hv capacitor (FIG. 7-26). Disconnect all leads from one side of the capacitor. Set the ohmmeter scale to R×10k and apply the test lead to the capacitor. Watch the meter hand charge and discharge. Reverse the procedure. A normal capacitor should never read under 100 kilohms.

7-26 To check the hv circuits, measure the resistance of each component. Suspect a defective magnetron when all hv components measure high with resistance tests.

Check the hv diode with the ohmmeter. Disconnect one terminal so the resistance reading is accurate. You should be able to have a reading in one direction and infinite resistance with reverse test leads and a normal diode. A normal diode may have a measurement of 150 kilohms. If the diode resistance is normal, measure the resistance from the heater terminal of the magnetron to the ground. Suspect a leaky magnetron if the ohmmeter reading is below 100k ohms with all heater wires pulled from the tube.

An open or poor heater terminal may prevent the magnetron from oscillating. Disconnect both heater terminals from the magnetron. Measure the resistance across these two cable ends. Most meters will read 0 ohms while a digital vom may indicate .001 ohms with a normal filament winding. Very seldom does the filament or heater winding go open since the coil is made up of heavy copper wire.

Go a step further and measure the ac voltage across the heater terminals. Remove the heater cables from the magnetron. If not excessive, high voltage may be found on the heater terminals of the magnetron. Clip the ac vom to the two heater cables of the power transformer. Keep the meter away from the metal oven area. Fire up the oven and measure the ac filament voltage. A normal filament reading is between 2.8 and 3.6 Vac. Although checking the hv circuits with the ohmmeter may take a few minutes longer, it is the safest method. Always be extremely careful while working around the hv circuits to prevent shock and injury.

Panasonic basic microwave oven cavity construction   The microwave oven cavity is a multimade cavity resonator, which is designed to resonate the microwaves emitted from the magnetron. A stainless steel, aluminum, or painted steel plate is used for the oven cavity.

The microwaves have several typical characteristics. They may be either absorbed, reflected, or transmitted, depending on the materials. The oven cavity is made with metallic walls that reflect microwaves (FIG. 7-27).

7-27   The rf energy radiates from the magnetron antenna, passes through the waveguide, and then is stirred up and moved around with the antenna blade over the food in the oven cavity (courtesy of Matsushita Electric Corporation of America).

Microwaves are emitted from the antenna of the magnetron and are transferred to the oven cavity through the waveguide, which is designed to transfer the microwaves to the oven cavity without any loss of the microwave energy. When the microwaves are fed into the oven cavity, they are either absorbed by the food or bounced around the oven cavity wall until they are eventually absorbed by the food. This difference in the energy mode will result in hot or cold spots.

In order to prevent the energy mode differences, microwave ovens may have a fan-like blade, called a *stirrer blade*, at the top of the oven cavity, or a turntable is provided on the bottom of the oven cavity to rotate the food. These devices prevent uneven cooking of the food by changing the energy path.

If the oven is operated empty or with large metal utensils inside, most of the microwaves are reflected back to the magnetron where they will dissipate as heat. This excess heat may damage the magnetron or shorten its operating life. Although technological advances in the magnetron tube and in oven designs have reduced the likelihood of a catastrophic failure, this problem is the reason for the warnings in the customer instruction books against the use of large metal utensils or the empty operation of oven.

MICROWAVE OVEN HOT SPOTS

Excessive arcing and hot spots found in the microwave oven may result from many different areas. Sparks or arcing occurring in the oven cavity may be caused by metal objects, bowls with metallic paint, or a temperature probe left lying in the oven or improperly inserted into the jack. Hot

7-28 Arcing may occur around the antenna, embossed ring, cracked antenna glass, open filament, and defective magnetron (courtesy of Matsushita Electric Corporation of America).

spots with the magnetron operation may result from not replacing the embossed ring or mesh gasket around the neck of the antenna assembly when replacing the magnetron, when the radiating stirrer blade is touching or close to the oven cavity, from a loose or damaged waveguide assembly, or a defective magnetron firing from the antenna to the oven ground (FIG. 7-28). Internal firing in the magnetron may be caused by an open filament, cracked antenna glass, or rf capacitors built right inside the magnetron. The magnetron must be replaced when arcing is found inside the tube.

Chapter 8

Locating
and replacing
the magnetron

Magnetrons produce microwaves through interaction of strong electric and magnetic fields (FIG. 8-1). When power is supplied to the magnetron from the high-voltage (hv) transformer, electrons are emitted from the cathode. They are accelerated toward the anode's 4,000 volt potential. A magnetic field along the tube axis forces electrons into a rotating spoked-wheel pattern before they reach the anode. As electrons pass the anode cavity openings, resonant cavities produce electromagnetic oscillations. These microwave oscillations are trapped and radiated by an antenna (FIG. 8-2). The microwave electric field reverses itself 2,450 million times a second. Polarized food or water molecules (separated electric charges) are vibrated back and forth in this field and this tremendous molecular friction produces heat.

Several different tests should be made to determine if the magnetron is defective or if the trouble exists in other circuits. Insufficient or no high voltage at the magnetron may be caused by a leaky magnetron or defective hv circuits. A leaky or shorted magnetron may lower the high voltage (FIG. 8-3). Sometimes the shorted magnetron will keep blowing the 15 amp fuse. Improper high voltage or a defective magnetron produces a no heat and no cook symptom.

HOW TO TEST

A current and hv test will quickly determine if the hv and magnetron circuits are normal. Correct high voltage at the heater or filament terminals of the magnetron may indicate the voltage-doubler circuits are functioning. High voltage may be monitored at the high side (negative) of the hv diode (FIG. 8-4). Be very careful—you are measuring up to 4,500 Vdc in

8-1   The anode of the magnetron is a hollow piece of iron with open cavities that are constructed of an even number of anode vanes pointing toward the filament from its outside wall (courtesy of Matsushita Electric Corporation of America).

8-2   The internal parts found in the typical magnetron tube. Notice a magnet is at both the top and bottom areas of the tube construction (courtesy of Matsushita Electric Corporation of America).

some microwave ovens. This voltage is negative with respect to the chassis as the cathode terminal of the diode is connected at ground potential. The high voltage may be measured with an hv dc voltmeter, high voltage probe, or a Magnameter.

When using a regular hv probe (found in the TV shop), the ground clips of the probe must be connected at the negative side of the hv diode.

8-3 A leaky or shorted magnetron may lower the high voltage. The old magnetron is shown on the left with the new replacement to the right in a Norelco microwave oven.

8-4 High voltage may be monitored at the high side (negative terminal of the hv side). Often, the heater or filament terminals are covered with plastic sleeves, and it is sometimes difficult to clip the voltmeter into the circuit.

The probe tip must be clipped to the chassis ground. Lay the probe on a book or manual and clip test leads to the probe tips. Do not hold the hv probe in your hands. Although the hv probe may not give a very accurate indication, at least you know high voltage is present.

An hv voltmeter or vtvm with an hv probe gives an accurate voltage reading, since the correct voltage scale and polarity are available on the vtvm. Again, clip the meter into the circuit with test leads. Keep the voltmeter insulated from the metal oven area. Place the voltmeter on a service manual or book. Always discharge the hv capacitor before attempting to take any voltage measurements. Low-voltage measurements at the filament terminals and hv diode may indicate trouble in the hv circuits or a leaky magnetron. Excessive high voltage may indicate that the magnetron is open.

Of course, a new oven test instrument called the Magnameter is ideal to check voltage and current measurements within the magnetron circuit (FIG. 8-5). A correct negative voltage at the hv diode indicates the hv circuits are normal. Simply flip the toggle switch to the low reading and measure the current pulled by the magnetron. No current reading indicates the magnetron is open. Lower current than normal may indicate a low emission tube. Higher than normal current measurements may indicate a leaky magnetron.

8-5 Here the GC Magnameter is connected to the magnetron circuit. With a flip of the switch, you may read voltage or current. This particular oven measures almost 2KV while operating.

CHECKING THE MAGNETRON WITHOUT AN HV METER

You can check the operation of a magnetron with a regular pocket vom and a 10 ohm resistor. Select a 10 watt and 10 ohm resistor and connect alligator clips to each end. Pull the power cord and discharge the capacitor before attaching the resistor. Remove the cathode or ground end of the hv diode. This positive end will be soldered or bolted to the metal chassis. Insert the 10 watt resistor in series with the diode (FIG. 8-6). Connect one end to the diode and the other to chassis ground. Make sure these connections are firm and tight. You may already find a 10 ohm resistor in older microwave ovens.

8-6 You can check the operation of the magnetron by inserting a 10 ohm 10 watt resistor in series with the cathode end of the hv diode. Remove the diode end from chassis ground and insert the resistor in series. Measure the voltage across the 10 ohm resistor.

Clip the negative lead of the vom to the chassis and the positive lead to the top side of the resistor, next to the diode. Switch the meter to the 100 Vdc scale. Prepare the oven for a cook test. Always start at or a little higher dc voltage scale. If the voltage is lower, go down to the next meter range. You prevent the meter from hitting the peg with this method.

With this setup, you are actually measuring the voltage across the 10 ohm resistor. If no voltage measurement is noted, you may assume no high voltage is present at the magnetron or the tube is defective since no current goes through the 10 ohm resistor. If there is a very high voltage drop across the resistor, this indicates the magnetron is pulling excessive current and is running extremely hot. A low-voltage reading indicates the magnetron is operating.

In a particular Sharp Model R-9314, a 10 ohm resistor was placed in series with the hv diode. With correctly applied high voltage and normal oven cooking conditions, the voltage measured across the 10 ohm resistor was 2.75 Vdc. The high voltage measured at the hv diode was 1,800 volts with the magnetron pulling 300 mills of current. You may assume in most ovens with a 10 ohm series connected resistor, the voltage will vary from 2.5 to 5 volts under normal operating conditions. A simple voltage test with the low-voltage vom may indicate the oven is functioning properly. These test voltages should be marked on the schematic for future reference.

CHECKING THE MAGNETRON WITH THE OHMMETER

Several tests of the magnetron may be made with the ohmmeter. Check for low resistance between heater terminals with no reading to chassis ground (FIG. 8-7). You should measure infinite resistance or no reading at all to ground. Use the megohm scale for leakage tests. If a very low reading is obtained, suspect a leaky magnetron. Remove all connecting cables from the heater terminals for this test. You may find a reading from one side of the heater wires to chassis indicating a leaky hv diode. Suspect an internal resistor in the diode case if the resistance is 10 megohms (FIG. 8-8). Some newer ovens have a bleeder resistor across the diode to bleed off the charge from the hv capacitor.

Filament
terminals

Magnetron
chassis
ground

Ohmmeter

8-7   You may be able to check the condition of the magnetron with the ohmmeter. Check for low resistance (under 1 ohm) between the heater terminals. An infinite reading may be obtained between the heater terminals and the magnetron chassis ground.

hv
capacitor

hv
transformer

.80 μF

hv diode
assembly

10 meg resistor

8-8   Disconnect the hv lead when making resistance measurements between the filament and ground terminals. If the resistance is 10 megohms, suspect a 10 megohm resistor built into the diode case. In a Sharp R9510 oven, a 10 megohm bleeder resistor is located inside the diode component.

Take another low ohm reading across the heater terminals of the magnetron. Usually, this reading is less than one ohm (FIG. 8-9). High resistance or no reading indicates the heater is open. A normal filament test made with the digital ohmmeter may show a fraction of 1 ohm. Always remove the filament transformer leads for this test or you may measure the resistance of the transformer winding and still have an open filament inside the magnetron. Although these two resistance readings help to locate a defective magnetron, you may still have a tube with weak emission or no high voltage applied to it.

8-9 Take another low-ohm reading across the heater terminals of the magnetron. Usually, this measurement is less than 1 ohm. A digital vom may measure a fraction of 1 ohm.

Norelco magnetron filament In normal oven cook operation, a visible glow from the heated tube filament can be seen through a small hole in the waveguide during a cook cycle.

If no filament glow is seen during a cook cycle, turn oven off, unplug power cord, and discharge capacitor with insulated screwdriver. Disconnect the hv leads from magnetron filament terminals. Measure the resistance across the two magnetron filament terminals with the ohmmeter set on R×1 (FIG. 8-10). If a normal resistance of less than 1 ohm between magnetron terminals is indicated on the meter, make a power transformer test. If high resistance or infinite resistance is indicated between magnetron terminals, replace the magnetron assembly.

Panasonic magnetron tests Continuity checks can only indicate an open filament or a shorted magnetron. To diagnose for an open filament or a shorted magnetron:

1. Isolate the magnetron from the circuit by disconnecting leads.
2. A continuity check across the magnetron filament terminals should indicate 1 ohm or less.
3. A continuity check between each filament terminal and the magnetron case should read open.

8-10 Checking the filament continuity with the DMM. Notice the resistance measurement is less than the vom.

CHECKING THE MAGNETRON WITH A COOKING TEST

The magnetron may be checked by simply performing a water temperature rise test. You will need a couple of 1 liter beakers, a glass thermometer, and a stopwatch. Before taking any tests, check for correct power-line voltage. Low power-line voltage will lower the magnetron output. This test should be made only with accurate test equipment.

Fill the two beakers with water and mark one (1) and the other (2). Stir the water in each beaker with the thermometer and record the temperature. Beaker temperature 1 is T1 and beaker 2 is T2. The average temperature of both beakers are as follows:

$$T = \frac{T1 + T2}{2}$$

Now, record the average reading.

Place both beakers in the center of the oven cavity. In the Sharp oven, place the beakers on the revolving glass tray. Set the oven for high power for only two minutes. Close the oven door and begin the cook cycle.

After two minutes are up and the oven has turned off, remove the two beakers. Stir the water with the thermometer and measure the rise in the temperature. Be careful and do this rather quickly. Record each beaker's temperature. Now take the average temperature as before. Subtract these two average temperatures. You will have the cold water temperature and now the warm water temperature. You should have a temperature rise of 15° to 20°F in a normal oven. If the temperature rise is below 8°F, the magnetron is cooking very slowly. When no temperature rise is noted, either the high voltage is low or there is a defective magnetron.

Now, check for proper line voltage. Check the high voltage at the magnetron with an hv test instrument. When the high voltage is present at the magnetron and no temperature rise is observed in the glass beakers, suspect a defective magnetron tube. Replace the magnetron tube and take another water test.

Panasonic's microwave output measurement   The power output of the magnetron can be determined by performing a simple water heating test. Check the line voltage under load. Low voltage will lower the magnetron output. Take the temperature readings and the heating time as accurately as possible. The equipment you will need includes two 1 liter beakers, a glass thermometer, and a wristwatch or stopwatch.

1. Fill each beaker with exactly one liter of tap water. Stir the water using the thermometer and record each beaker temperature (recorded as T1A and T1B).

2. Place both beakers on the center of the cook plate. Set the oven for HIGH power, and heat it for exactly two minutes.

3. Stir the water again and read the temperature of each beaker (recorded as T2A and T2B).

4. Obtain the temperature rise by the following formula:

   Average temperature rise =

   $$\frac{(T2A + T2B) - (T1A + T1B)}{2}$$

   The normal temperature rise for these models should be 16° to 20°F (8.9° to 11°C) at the HIGH power selection with the oven operating at the specified line voltage.

5. To obtain the power output, multiply the average temperature rise by 39 (or 70 if thermometer is in Celsius).

   EXAMPLE:
   If you find average temperature rise of 18°F, the oven's output is 18 × 39 = 702 watts. **NOTE:** If only one beaker is available, you may conduct the test by heating one liter of water for one minute. In this instance, you will find the rise in temperature is smaller than the above specifications due to the microwave absorption characteristic. You may judge the magnetron to be normal when the temperature rise is greater than 13.5°F (or 7.5°C).

MAGNETRON FAILURE

When you find the magnetron is arcing inside the top oven area, suspect a broken seal or that food particles are causing excessive arcing. Sometimes with poor handling and striking the glass around the antenna area may let air inside, destroying the vacuum tube. This may occur while installing a new magnetron or cracking can be caused by thermal and mechanical stress.

A suck-in may occur with the result of abnormally high power on the antenna glass, which softens the glass. The outside pressure pushes the glass inward toward the vacuum until a small hole is formed (FIG. 8-11). This may occur with improper use of the oven such as using metal cooking utensils in the oven and with an empty oven. Often the breakdown is at the antenna glass to the metal seal. You can definitely hear any arcing occurring inside the magnetron. After determining the magnetron is

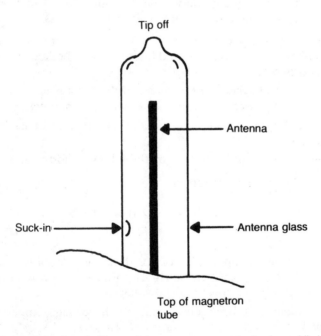

8-11   The top of the glass shell around the magnetron may be broken while installing or removing the tube. A suck-in at the bottom of the neck, caused by outside pressure may occur from using metal utensils inside the oven cavity or from running an empty oven.

defective, immediately shut off the oven to prevent hv component breakdown.

An open heater or filament may cause internal arcing. Discharge the hv capacitor, remove all cables to the heater terminals and take a continuity test between the heater terminals. Very seldom does the heater short internally. Rough handling or shipping may cause the filament to go open.

The magnetron tube may have low emission. This may occur after many years of operation. Often, the tube suffering from low emission may take longer to heat up. When you receive a complaint of slow cooking, suspect a defective magnetron. A tube with low emission may never reach the correct current requirements. The Magnameter is ideal for reading the current of any magnetron. Simply see what the current measurement is when turned on and if the current increases as the oven operates. It's nice to take voltage and current readings of each oven repaired. Mark these measurements right on the oven schematic for future reference.

Burns food

Suspect a defective magnetron when the complaint is that the oven runs hot and burns the food. Replace the magnetron if the oven appears extremely hot in just a few minutes. Often the thermal protector switch will intermittently shut down the oven when the magnetron appears too warm. Pull the power plug. Discharge the hv capacitor and touch the magnetron assembly. Replace the magnetron if it is too hot to touch. Take a current test and you will find if the hot magnetron is pulling excessive current.

Arcing in the oven

Excessive arcing inside the oven may be caused by a shorted or leaky magnetron. You may find the waveguide cover is burning. Remove the cover and fire the oven up. Sometimes grease behind the cover will cause arcing and burning of the cover. Replace the magnetron if the arcing continues. When excessive arcing occurs turn the oven off at once.

Dead (No cooking)

A defective magnetron may cause the no-cooking symptom. First, take a voltage and current measurement of the hv circuits. Higher than normal high voltage indicates an open tube. No high voltage may indicate problems in the hv or low-voltage circuits. Poor tube emission will definitely show very little current measurement. A magnetron with no emission may be caused by a defective cathode element. Discharge the hv capacitor. Take resistance measurements at the filament terminals. No resistance reading indicates an internal open filament.

Erratic or slow cooking

A defective magnetron may produce erratic or slow cooking symptoms. With high voltage and current monitoring you can quickly determine if the magnetron or other circuits are not functioning. The hv and current measurement should be read within two or three seconds. If the current reading is erratic and the high voltage fairly normal, you may assume the magnetron is defective. Intermittent high voltage may indicate a defective hv circuit or magnetron. In some extreme cases, replacing the magnetron may be the only solution. Slow cooking may be caused by poor heater connections at the clip-on terminals. Check for burned and overheated connections.

Everything operates—No heat—No cooking

This is a very common symptom in most microwave ovens. All lights seem to be on and the turntable is rotating, but there is no cooking inside the oven. Check with a water cook test. First, determine if voltage is applied to the magnetron. Check the magnetron current. When correct voltage is found at the heater terminals to ground, suspect a defective magnetron. No current reading indicates a defective magnetron. Check the heater continuity and resistance measurement between heater and chassis ground to determine if the magnetron is defective, when an hv voltmeter and current meter are not available.

Cuts off after a few minutes

When the magnetron becomes quite warm after several minutes of operation, the thermal protector switch may open, removing the power-line voltage from the primary winding of the hv transformer. After the thermal protector cools down, the oven begins to cook once again (FIG. 8-12). You may monitor the intermittent cooking with the ac voltmeter connected across the thermal switch. A pigtail bulb may be used here as a monitor device (FIG. 8-13). When the bulb lights or the power-line voltage is indicated on the ac meter, the oven is not cooking. Also, you may hear a loading-down noise from the power transformer under a cooking load, and when the voltage is disconnected the noise disappears.

Check the terminals of the thermal protector switch for burned areas. A defective thermal switch may cause the same symptoms. Replace the thermal switch if the magnetron is not running very warm. If the new switch is normal and the magnetron appears quite warm, install a new magnetron tube.

Norelco magnetron thermal protector   The magnetron thermal protector, located on the magnetron assembly, is designed to prevent damage to the magnetron if an overheated condition develops in the tube due to fan failure, obstructed air ducts, etc. (FIG. 8-14). The thermal protector opens at approximately 300 °F and resets automatically at approximately 200 °F.

Under normal operating conditions, the thermal protector remains closed. However, if abnormally high temperatures within the magnetron approach a critical level, the thermal protector will interrupt the circuit to the magnetron and coil of the surge relay. When the magnetron has

8-12   A defective thermal protector switch may cause erratic or slow cooking. When the magnetron appears warm, the thermal switch opens up, shutting the high-voltage down. After the magnetron cools down, the thermal switch applies voltage to the magnetron circuit.

8-13   Use a pigtail bulb clipped across the thermal protector switch for a monitor. When the light comes on voltage is not applied to the magnetron. Voltage is applied when the 100 watt lamp goes out.

8-14   The location of the thermal cutout upon the magnetron tube of a Norelco microwave oven (courtesy of Norelco Corp.).

cooled to a safe operating temperature, the thermal protector closes and·a
cook cycle can be resumed.

INTERMITTENT THERMAL UNIT

It's possible to locate an erratic or intermittent thermal cutout when the
oven cooks intermittently. Often, the thermal switch element is weak or
has burned contacts. Monitor the thermal cutout with an ac voltmeter or a
100 watt light bulb (FIG. 8-15). Under these conditions, check to see if the
magnetron is overheating and quite warm. If the magnetron is fairly cool
and the cutout is erratic in operation, replace it. Do not try to repair the
thermal cutout component.

REMOVING THE MAGNETRON

Before attempting to remove the magnetron tube, once again discharge
the hv capacitor. Take a visual inspection of the components surrounding
the magnetron. You may have to remove a blower motor assembly before
getting at the tube (FIG. 8-16). Disconnect wire leads from the thermal pro-

8-15 Checking the thermal cutout with the ac voltmeter of a DMM.

8-16 You may have to remove the blower motor before removing the magnetron. Here the fan motor is connected to the bottom side of the magnetron.

tector switch and heater terminals. Be careful when removing the wires so you do not damage other components. Mark down where the various cables are connected.

After the magnetron appears free of wires and shields, remove the four mounting nuts holding the magnetron to the waveguide (FIG. 8-17).

8-17 Visually inspect all components surrounding the magnetron before attempting to remove it. Here, in a Norelco MCS 8100 model, the blower motor, metal brackets, and triac module must be removed before removing the magnetron (courtesy of Norelco Corp.).

Use socket wrenches to remove these nuts instead of pliers. Be *very careful* not to let the magnet pull the tool from your hands and break the glass seal of the magnetron. These magnets are very strong and may grab the socket wrench. Just loosen the nuts—then remove by hand while holding the bottom of the magnetron so it will not drop. The magnetron assembly is quite heavy.

Now lower the magnetron until the tube is clear of the waveguide assembly. You may have to tip the tube to remove it from the oven. Be careful not to break the glass antenna assembly. This is located on top of the magnetron. When mounted, the antenna assembly protrudes through the waveguide assembly. Place the magnetron on a firm surface and out of the way to prevent breakage. Here is a 10-step manufacturer's instruction for removing the magnetron:

1. Remove the blower motor assembly.
2. Remove the exhaust duct.
3. Remove the thermal protector switch assembly (two metal screws).
4. Remove the cable leads from the heater terminals (mark if needed).

5. Remove all metal flange or components that may be in the way of the magnetron.

6. Remove the four mounting nuts holding the magnetron to the waveguide assembly (be very careful with tools around the magnetron).

7. Hold the bottom of the magnetron so it will not drop and damage the other components.

8. Lower the magnetron until the tube is clear of the waveguide assembly.

9. Remove the tube assembly completely from the oven.

10. Check to see if the rf gasket is still on this tube. This gasket may be needed for the new tube.

KEEP METAL TOOLS AWAY

When working around the magnetron with screwdrivers, pliers, and socket wrenches, keep the tools away from the glass antenna assembly. This unit sticks above the magnetron and can easily be broken with the strong magnets pulling a tool out of your hands. Be very careful when removing the magnetron mounting screws or nuts and replacing them.

REPLACING THE TUBE

The magnetron should be replaced in the reverse order of removal. Make sure the new magnetron has the same part number or it may be a substitute recommended by the manufacturer. You may find several newer types are subbed for an earlier version. Inspect the rf gasket before installing (FIG. 8-18). If the new tube does not have one, use the old rf gasket. Replace the gasket with a new one when excessive arcing has occurred around the tube or the gasket is damaged.

It's possible to replace one magnetron with one from another manufacturer (FIG. 8-19). You will find some tubes are common in several ovens. In fact, the part numbers on the magnetron may be the same. Of course, do not sub a different type of magnetron. You may find that some magnetrons are larger than others and mount differently. The heater terminals may stick out at the wrong angle for correct mounting. Follow the manufacturer's replacement procedure.

Be sure to inspect the new tube before installation. Check for damage marks on the shipping box. When the bottom of the tube area is pushed to one side or up at an angle, the magnetron may arc internally. The glass sealed area may be cracked during shipment. All cooling fins should be equally spaced and not crushed.

Double check all mounting screws. Check the schematic or installation notes for correct hookup. Make sure the heater cables are in place. If the clips fit over the heater terminals rather easily, pinch the area together with pliers to make a greater contact. These contact clips will arc and burn if there is a poor contact, causing intermittent or erratic oven opera-

MAGNETRON
MOUNTING NUTS

MOUNTING SCREWS
BLOWER BRACKET

RF GASKET

THERMAL
PROTECTOR

MOUNTING
SCREWS

MAGNETRON TUBE

MOUNTING
NUTS

BLOWER MOTOR

8-18 Inspect the rf gasket before installation. Use the old one if it is not damaged. Replace the gasket when excessive arcing has occurred around the tube or when the gasket is dented.

8-19 Sometimes you can replace the magnetron tube with one from another manufacturer's magnetron.

tion. Before replacing the back cover, check for leakage around the magnetron area and a normal cooking test.

Panasonic magnetron replacement The positive lock connector is a specially designed locking connector that you will find in many lead wire connections. To remove this connector, pull the lead wire by pressing an extruded lever in the center of receptacle terminal.

1. Discharge the hv capacitor.
2. Remove two screws holding magnetron thermal cutout.
3. Remove three screws holding air guide and detach air guide.
4. Disconnect two hv lead wires from the magnetron filament terminals.
5. Remove four screws holding the magnetron. After replacement of the magnetron, tighten mounting screws properly making sure there is no gap between waveguide and the magnetron to prevent microwave leakage.

COSTS TOO MUCH TO FIX

Always provide a repair estimate for any repair involving magnetron replacement. These tubes are fairly expensive and, when added to normal labor costs, the repair could cost more than buying a new oven. When you provide a cost estimate, customers can make up their own minds whether to have the oven repaired or not. After making sure the magnetron is defective through hv, current, resistance, and cooking tests, always provide an itemized estimate for the customer.

TOO HOT TO HANDLE

One must be careful when working around microwave ovens to prevent burns from steamy hot dishes and food. Any time heat or cooking is involved, it is very easy for a person to receive burns on the hands and face. Although most burns are from the cooked food or utensils, most people are afraid of getting burned or injured with radiation from the oven. It is impossible to have rf radiation if the oven has been checked for leakage with an approved leakage tester.

Excessive grease found in the oven cavity may cause fire or burning of plastic covers and waveguide covers. Often greasy food (such as bacon) accumulates after several years. The grease may run down behind the plastic shelf guide or behind the waveguide covers. Although the plastic waveguide covers are up at the top, they still collect grease on the top side. A lot of oven maintenance may be avoided if the plastic shelving and waveguide covers are removed and cleaned with a mild detergent.

In some early oven models, plastic shelf guides are found on each side of the oven cavity. These are held in place with metal screws. The grease collects behind the plastic and metal screws, causing the rf energy to burn the plastic guide assembly. Simply removing the plastic guide and washing it may prevent an oven fire and costly maintenance problem.

Burning of waveguide covers may be caused by excessive grease collected on top of the cover and by a defective magnetron. Often the defective magnetron may burn only a small section of the waveguide cover. Excessive grease on the cover may start to burn in several areas. Always replace the waveguide cover when it is burned or when replacing a defective magnetron. You may not locate a waveguide cover in the early microwave ovens.

TWO EXPLOSIVE EXAMPLES

In this early microwave oven, only a round hole is found in the aluminum hole cover (FIG. 8-20). Undoubtedly, the owner was cooking whole eggs in the oven. Eggs may be cooked in this manner if a couple of pin size holes are punched in the top of the shell. In this case, part of the egg and shell exploded going through the 1 inch diameter hole and lodging against the magnetron. The magnetron began to arc between antenna and gasket area. The longer the oven cooked the greater the arcing of the magnetron tube.

The magnetron was removed and inspected. In this particular case, the magnetron was not damaged. But, if the customer continued to cook or allow the oven to arc for several minutes, the magnetron may have become damaged, resulting in a very expensive repair job. Simply cleaning off the magnetron antenna and cavity area solved the oven arcing problem. Again, careful cooking methods must be observed when operating any microwave oven.

Some foods may become explosive dishes when cooked in microwave ovens. Most oven manufacturers warn that popcorn must be popped only in ready-made popcorn bags. In this particular oven, popcorn was placed in a brown paper bag without any steam holes, resulting in an open fire within the oven cavity. The paper bag caught on fire, melting down the top plastic waveguide cover (FIG. 8-21). Of course, this plastic cover is inexpensive and easy to replace. If the fire had not been extinguished in time, the whole plastic front cover would have been burned. The inside plastic liner was burned, showing a burned area in the front door area. The front door assembly was removed for inspection. Here only the inside plastic area was burned. If the choked gaskets and whole door assembly had to be replaced, it would have resulted in a very expensive repair job. Always inspect the door area for possible choke damage and double check when taking leakage tests.

HOW TO ORDER A NEW TUBE

The new magnetron should be ordered from the manufacturer or manufacturer's distributor, if you are a dealer or service dealer. Most manufacturers will refer you to their various service centers. The new tube may be ordered from the dealer or microwave oven service center if you are

8-20   Excessive arcing was noticed in a microwave oven. Removing the magnetron and waveguide cover showed sections of egg shells. Undoubtedly, the egg shells went through the small, round hold and caused the magnetron to arc over.

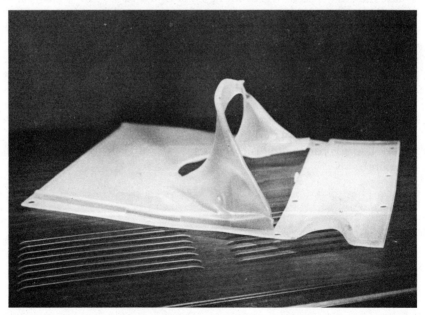

8-21   Here a plastic waveguide cooking cover was melted down when popcorn was placed in a brown paper bag. The manufacturer's warnings on how to cook food in the microwave oven should be followed.

going to install the magnetron yourself. You may find the magnetron in stock if it is a popular model. Otherwise, it may take from three days to several weeks to obtain a replacement.

If handy, check the service manual for the correct magnetron part number. This same number may be found on the tube. In case the service manual is not available, use the numbers found on the magnetron. Don't forget to give both tube part number and the model number of the microwave oven when ordering. When the tube arrives, the manufacturer or distributor may have subbed a new magnetron. The new tube will mount in the same place as the old one.

Check the tube over carefully when it arrives. Notice if the magnetron is packed in two separate boxes. Most manufacturers or parts distributors will ship the magnetron in a regular parts box, placed in a larger one for safe shipment. Check the boxes for broken areas. If one side of the metal flange assembly of the magnetron is bent out of line, refuse to install the tube. You may find it will arc internally when fired up. Return the defective magnetron.

If the new tube appears intact, install the thermal protector switch in the matching screw holes. A new switch should be installed if it is defective or appears to have been operating quite warm. Now, the magnetron is ready to be mounted in the waveguide assembly.

HOW TO REPLACE WHILE STILL UNDER WARRANTY

In case the old magnetron is in warranty, pack the tube in the same carton the new one came in. Most magnetrons are warranted for a period of five or seven years. Some manufacturers issue a registration number when the oven is sold. Either the owner or dealer should register the oven with the manufacturer. If the oven has not been registered, a bill of sale must be attached to the warranty repair tag. Most oven manufacturers have their own warranty forms or will acknowledge NRA form number 317-515.

Make sure the magnetron is packed in its original carton plus another one. Pack the tube well to prevent breakage. The warranty and work order form should be placed in the same carton. Fill out the forms completely. Usually, parts credit is issued and warranty labor paid the following month.

Chapter 9

Servicing oven motors

You may find there are up to seven different motor operations in a microwave oven. The fan or blower keeps the magnetron cool and moves the air in and out of the various vented areas. You may find two separate fan motors in some ovens. The cooking time may be controlled by a timer-clock motor. A stirrer motor spreads the rf energy out over the food in the oven cavity. Now, in many ovens you may locate a turntable motor, rotating the food for even cooking (FIG. 9-1). A vari-motor may be used for intermittent cooking or defrosting the frozen food before actual cooking begins. In the convection oven, you may have a heater or auxiliary blower motor to circulate the heat provided by the heating elements. The convection oven may also have a small motor in the convection timer.

The rotation of each different motor may help you locate a defective component in the oven. You may use each motor tied in a different leg of the schematic diagram for trouble indications (FIG. 9-2). When the fan or blower motor does not operate, you know the defective component is in the input of the power-line circuits. No rotation of the turntable or stirrer motor may indicate a defective oven relay or contacts, since these motors are connected into the circuit after the relays. A rotation of the blower motor connected across the primary winding of the power transformer indicates the low-voltage stages are normal.

Although the motors found in the microwave oven cause many different problems, a defective motor is easy to test and locate. A defective motor may be located by voltage and continuity tests. With the oven in operation, measure the ac voltage across the motor terminals (FIG. 9-3). Remove the power plug and discharge the hv capacitor. Set the ohmmeter to the R×1 scale. Measure the resistance across the motor terminals. An

NN-8507/NN-8807/NN-8907 CPH

NOTE: Door is closed.
Unit is not operated.
⏚ Ground
⫫ Chassis

IMPORTANT SAFETY NOTE: THE SHADED AREA ON THIS SCHEMATIC DIAGRAM INCORPORATES SPECIAL FEATURES IMPORTANT FOR PROTECTION FROM MICROWAVE RADIATION, FIRE AND ELECTRICAL SHOCK HAZARDS. WHEN SERVICING, IT IS ESSENTIAL THAT ONLY MANUFACTURER'S SPECIFIED PARTS BE USED FOR THE CRITICAL COMPONENTS IN THE SHADED AREAS OF THE SCHEMATIC DIAGRAM.

9-1 A typical oven schematic showing where the fan blower and turntable motor are used in the microwave oven circuit. Both motors are controlled by the primary latch switch (courtesy of Matsushita Electric Corporation of America).

9-2 The various motors may be used as indicators in low-voltage circuits. When the fan or blower does not rotate, the defective component is in the input of the power-line circuit.

9-3 A defective motor can be located by voltage and continuity tests. Set the ohmmeter to the R x 1 scale for a motor field continuity test.

open field will have no ohmmeter reading. Very low ohmmeter measurements may indicate a shorted field winding.

Just clip the lightbulb pigtail test light across the fan or blower motor terminals to make sure power is applied.

FAN MOTOR PROBLEMS

The fan or blower motor problems may produce many different symptoms in the microwave oven (FIG. 9-4). A dead or nonrotating blower motor may cause the magnetron to overheat, causing an intermittent cooking shut-down. When all other components seem to be in operation and the fan is not rotating, check the ac voltage across the motor termi-

9-4 A dead fan motor might cause the magnetron to overheat. In this early Goldstar microwave oven, the fan motor is bolted right to the magnetron assembly. Notice the dirt on the fan blades.

nals. If the power-line voltage is present, pull the power plug and remove one motor terminal wire. Take a resistance measurement across the motor field terminals. An open field measurement may be caused by a poor or open motor socket connection. A quick method to see if ac voltage is applied to the motor field is to place a screwdriver blade against the metal motor area and the blade should vibrate.

Intermittent fan blower operation may be caused by a poor contact of the motor socket terminals. Sometimes these sockets will vibrate loose and make a poor contact (FIG. 9-5). Check for poor or old cable wires connecting to the motor field coils. Use the R×1 ohmmeter range for all motor continuity tests. (The resistance of a fan motor in a Sharp R8310 oven was found to be 41 ohms with a digital vom.)

An overheated motor may produce a shut-down after several hours of operation. Usually, the field coil shorts the turns within the motor winding when the motor overheats. When the fan blower to the magnetron quits rotating, the magnetron becomes overheated and the thermal-switch cuts off power to the magnetron. Most fan or blower motors operating at the correct speed produce quite a rush of air inside the oven. Compare the resistance reading of the motor field with that given by the manufacturer. A digital vom takes a very accurate measurement of all motor field coils. Feel the motor body and if it is very warm then replace it.

When moisture appears around the door area, suspect poor ventilation. Check the fan motor for correct rotation. Clean out all vent areas.

9-5 Intermittent fan operation may be caused by poor contact of the motor socket terminals. Flex the cable and socket while the motor is operating to locate the poor connection. This fan brings in outside air from a vented area and blows against the magnetron for cooling.

Don't overlook the clogged fin areas of the magnetron tube. Improper intake of cool air and venting of hot air may cause the magnetron to overheat. Check for correct vent and damper action especially in the convection type ovens in the fan blower area. Also, check for enough air space around the sides of the oven vent areas.

A noisy fan blower motor may be produced by dry bearings or loose motor mounting screws. Check the fan blade if it is bent out of line or loose on the motor shaft. A bent fan blade may cause the motor to vibrate producing a noise inside the oven area. Often a high-pitched noise is a sign of dry motor bearings. Gummed up motor bearings may cause the motor to overheat and run slowly. If the blade cannot be straightened properly, then replace it. Most fan motors are adequately lubricated. Don't overlook a dry fan motor bearing in the older ovens. Light motor oil placed on the motor bearings may solve the noisy motor bearing problem.

FAN STARTS AFTER 10 MINUTES

Suspect a dry or gummed up fan motor when the blower does not come on for 5 or 10 minutes (FIG. 9-6). This condition may exist if the oven is 5 to 10 years old. The fan blower motor may have gummed up bearings caused by excess dirt and grease drawn into the fan area. Sometimes the thermal protector may shut down the magnetron if the fan motor takes a

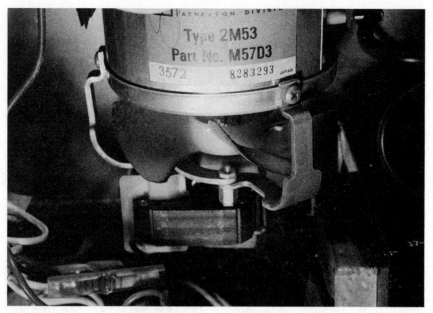

9-6 Suspect a dry or gummed up bearing of the fan motor when the blower does not come on for 5 or 10 minutes.

long time to come on. Clean up and lubricate the fan motor when this condition exists. If you don't, the magnetron tube may be damaged with erratic cooking problems.

THE BLOWER MOTOR

The fan or blower motor rotates a blade that draws cool air from outside the oven. This cool air may be directed through different air vanes surrounding the magnetron. In some ovens, the fan motor is bolted to the bottom side of the magnetron. You may find two different fans in some ovens, one moving the air in and out with the other fan cooling down the magnetron tube. Most of the air is exhausted directly through the back vents. In some ovens, the air is channeled through the cavity to remove steam or vapors given off by the food. Often the fan motor operates during microwave and convection cooking in the combination type oven (FIG. 9-7).

The fan blade may be made of plastic or light metal. The blade may be held in place by an allen set-screw or a nut mounted on the outside of the blade area. A bent fan blade may cause the fan to vibrate creating a vibrating noise. Some fan motors are mounted in rubber mounts for quiet operation.

In some fan or blower motors, the motor coil may be fused internally. The fuse is embedded in the motor winding. If the fuse opens, the motor must be replaced. The motor field coil resistance may vary from 10 to 50 ohms. In the Hardwick ovens, the blower motor resistance may be from

9-7 The fan motor operates during microwave and convection cooking in a combination oven. Here the fan motor runs all the time and the convection motor rotates only when the switch is in the convection cooking mode.

16 to 22 ohms, while in the Amana ovens, the blower fan coil resistance may be from 5 to 10 ohms. The blower motor resistance in a small wattage Goldstar microwave oven is 17 ohms (FIG. 9-8).

Remove one lead from the motor for accurate ohmmeter measurements. In some ovens, the fan motor connections unplug and you may find the leads wired directly into the motor. Infinite resistance is measured from the metal motor frame to one lead of the field coil (FIG. 9-9). Any resistance between the motor frame and field lead indicates a breakdown in the motor coil winding. Replace the blower motor to prevent shock hazard to the operator. Remember, the fan blower may also rotate the stirrer fan or blade with a pulley and belt arrangement. A typical fan motor replacement procedure is as follows:

1. Remove the belt between the gear box and the fan motor pulley (only with combination fan and stirrer motor).
2. Disconnect the two-wire plug or wire connectors of the fan motor.
3. Remove all screws holding the fan motor bracket to the oven.
4. If the motor is fastened to a separate bracket, remove these bolts and nuts.
5. Remove the fan blade (only if the motor cannot be removed).
6. Remove the ground strap.
7. On a combination stirrer-and-fan motor, remove the pulley to place it on the new motor.
8. Install the new fan motor by reversing the removal procedure.

Norelco RR7000 fan motor replacement Disconnect the two leads

9-8   Here the blower motor resistance in a small Goldstar microwave oven is .001 ohms with the DMM.

9-9   Infinite resistance should be noted between the metal motor frame and either side of the motor terminals. Any ohmmeter reading may indicate a leaky or shorted internal coil winding to ground.

from the fan motor. Remove the four blower assembly mounting screws secured to the air baffle (FIG. 9-10). Remove the blower motor. Remove the blower fan by pulling the fan from the motor shaft. The blower mounting brackets can be removed by taking out two mounting nuts and bolts that

9-10   In a Norelco RR7000 model, the fan motor can be dismounted by removing the four blower assembly screws (courtesy of Norelco Corp.).

mount the brackets to the blower motor. To install a new motor, reverse the above procedure.

THE STIRRER MOTOR

The stirrer motor rotates or circulates the rf energy emitted from the magnetron waveguide assembly. The stirrer motor assembly provides even cooking and eliminates dead spots in the oven. In some Sharp microwave ovens, the food is turned instead of using a stirrer motor. A stirrer motor assembly is located at the top side of the oven. In some ovens, you may find a pulley with a long flat belt driving the stirrer blade from the blower fan assembly (FIG. 9-11). Here the fan motor serves two functions, circulating the cool air and rotating the stirrer blade. Check for a broken belt when the stirrer blade is not rotating. A nonrotating stirrer motor may cause dead spots and improper cooking.

Norelco stirrer motor   The stirrer motor is mounted on a bracket and operates a separate stirrer pulley via a belt. The blade revolves between 55 and 60rpm and reflects electromagnetic energy produced by the magne-

9-11   The center for blade operation in a convection Sharp Model R8310 is rotated by the fan blower motor. The fan motor circulates the air cooling, and a pulley shaft rotates the convection fan assembly.

tron tube, back and forth, between the sides, top, and bottom of the metal cooking cavity (FIG. 9-12). This allows rf energy to penetrate food from all sides and gives a uniform heating pattern.

Since the stirrer motor has a slower rotation, the field coil resistance may be from 300 to 3 kilohms. Check the manufacturer's literature when you suspect a defective stirrer motor. In a Norelco MCS 6100 model, the stirrer motor resistance is 2.7k, while in a Hardwick EN 228 model, the motor resistance is 450 ohms. Check the continuity of the motor terminals with the low-ohm scale of the vom.

Noise caused by the stirrer motor may originate when the blade becomes loose on the motor shaft. The blade may rub or hit against the air chamber assembly. Check for a noisy pulley assembly at the hub or where the long belt goes over the oven edge (FIG. 9-13). This belt may be a flat type with serrated teeth. Sometimes the pulleys may need lubrication or need to be tightened. Another source of noise may be caused by a dry stirrer shaft bearing.

The stirrer motor or blade assembly may be difficult to remove. Usually the stirrer fan blade is located between the large waveguide cover and air guide assembly. In most ovens, the blade or motor must be removed from the top side of the oven. Remove all waveguide metal screws holding the stirrer blade assembly. Some parts of the exhaust and air duct assembly must be removed before the stirrer assembly is free. Replace the stirrer motor and air duct assembly in the reverse order.

9-12 Both the fan and stirrer motor operate at the same time and are controlled by the control panel in a Norelco oven (courtesy of Norelco Corp.).

9-13 In a Quasar oven, the fan motor provides cooling and also rotates a gear box assembly. Notice the flat belt from the fan motor assembly. A long rubber belt drives the stirrer fan at the top area. Check for a noisy pulley or bearing at the various belt pulley assembly.

Panasonic stirrer motor replacement Lay the oven on its top side. Remove the two screws holding the motor cover (FIG. 9-14). In this oven, the stirrer motor and antenna assembly is located in the bottom of the oven cavity area. Disconnect the two lead wires from the turntable/antenna motor. Remove the two screws holding turntable/antenna motor (FIG. 9-15).

Norelco stirrer motor replacement Remove the stirrer belt. Remove the rubber pulley grommet by lifting off the shaft. Remove the two screws that secure the stirrer motor to the mounting bracket and remove the motor. In some cases, the wire must be cut and the replacement motor must be spliced and insulated with electrical tape or approved splice connectors. Reassemble the motor in reverse order.

The stirrer blade may be removed by stretching the belt above the rubber pulley grommets and lifting it off. Remove the shaft retainer fasteners by squeezing two tabs with needle-nose pliers and lifting them up. Remove the pulley from the shaft (FIG. 9-16), and remove the stirrer cover. The stirrer blade can be brought downward and out of the oven. Remove the stirrer blade by removing the two screws that hold the blade to the shaft if a blade replacement is needed.

9-14 Lay the oven on its top with padding underneath to prevent scratching the top side of the oven. Remove the bottom motor cover of a Panasonic MQ8877/8877 microwave oven (courtesy of Matsushita Electric Corporation of America).

9-15 Disconnect the two lead wires from the turntable/antenna motor and remove the two screws holding turntable (courtesy of Matsushita Electric Corporation of America).

THE TIMER MOTOR

The timer switch contacts are mechanically opened and closed by turning the dial knob located on the timer motor shaft. These contacts control the current path to the primary winding of the hv transformer and many other oven components. The correct cooking time is set on the timer unit and the timer motor begins to rotate. When the timer reaches the 0 point on the scale, the timer opens the circuit and the cook cycle stops (FIG. 9-17).

9-16 Remove the two tabs with needle-nose pliers and lift upward. Remove the two screws holding the turntable (courtesy of Norelco).

9-17 A simple timer motor assembly is found in a Litton manually-operated oven. A microswitch is used to switch in the timer of the low-voltage circuits.

The timer motor operates directly from the power line (120 Vac). Measure the power-line voltage across the motor terminal (FIG. 9-18). No ac voltage may be caused by defective interlocks or cook switches. If 120 Vac is measured across the motor terminals, and the motor doesn't operate, take a continuity test. Pull the power plug, discharge the hv capacitor and remove one motor terminal lead. Set the vom to the R×1 scale. Most timer motor resistance tests are from 100 to 150 ohms.

9-18 The timer motor may be checked with ac voltage and ohmmeter tests. Measure the power-line voltage across the timer motor terminals. If 120 Vac is measured, suspect a defective timer motor or gear assembly when the timer does not count down the time of cooking.

Replace the timer assembly if the motor winding is open or mechanically defective. A defective timer assembly may be erratic in operation and may never complete the correct cooking time. Sometimes the timer may stop without returning to 0.

Replace the defective timer assembly with one having the original part number. Although you may have to move several components out of the way, most timer assemblies are easy to remove and replace. Remove all mounting screws to free the timer assembly. Replace the new timer assembly. Remount all other components and reconnect the wiring cables. Check the timer operation by rotating the timer knob to the 10 minute mark, then return it to 0. The bell should ring on most timers (FIG. 9-19). If the timer pointer does not correctly come to the 0 position, adjust the position of the timer assembly with the mounting screws loose. Then tighten the mounting screws and check once again.

CONVECTION TIMER MOTOR

In some ovens, a convection timer motor controls the cooking time and damper rod. The timer contacts are mechanically closed or opened by

9-19 Most manual timer assemblies have a bell that will ring when the time is up. Check the timer by rotating the timer knob to the 10 minute mode, then back to 0 and check to see if the bell rings.

rotating the convection timing knob. The convection timer may cook up to two hours. The timer motor may be energized through the convection timer contacts. A cam on the timer shaft operates to close the damper so no hot air escapes from the oven cavity. The damper is opened when the timer shuts off, letting hot air out of the convection oven.

TURNTABLE MOTOR ASSEMBLY

The turntable motor rotates the food for even cooking and prevents hot spots. The food is placed in a glass tray, with the tray rotating around on small plastic rollers. This turntable starts to operate when the oven relay contacts are made. The turntable rotation may be used as an indicator in troubleshooting the oven circuits. You know all low-voltage circuits are normal when the turntable is rotating. You will find the turntable assembly only in some microwave ovens (FIG. 9-20).

The most common problems found with the turntable assembly are excessive noise while rotating, intermittent operation, and no rotation. Check the small plastic wheels found in the oven cavity for possible squeaky or dry bearing noises. Remove the glass turntable and let the motor rotate. If the noise has disappeared, lubricate the small wheels, holding up the glass tray. Only a drop of oil on each wheel will do. In case the noise is still present, check the plastic spindle bushing for dry or worn areas. You may hear a loud grinding noise with a broken or split spindle bushing. Check for a noisy and dry gear box assembly.

9-20 The turntable motor assembly is located in various Sharp, Panasonic, and Goldstar models. The turntable rotation may be used as an indicator in troubleshooting the low-voltage oven circuits.

Intermittent operation may be caused by a dry or defective plastic bushing (FIG. 9-21). Remove the turntable motor and inspect the bushing area. Check the bushing in operation. When found dry, lubricate the bushing with a light grease. Always replace a cracked or broken plastic bushing assembly. A jammed gear box may cause the turntable to run slow or have intermittent operation. Check for plenty of grease in the gear box assembly. Worn or broken gears may cause the gear box to freeze up and produce slow or intermittent rotation. Also, check for poor connections or improper voltage at the motor winding terminals for intermittent operation.

A dead turntable motor may be caused by no voltage (120 Vac) at the motor terminals, a frozen bushing, or a jammed gear box. Check the ac voltage at the motor terminals with the oven in operation. Usually, the motor assembly is normal when voltage is found at the field coil. If a vom is not handy, place a screwdriver blade against the motor frame and you should feel a magnetic pull or vibration on the motor assembly. When the motor will not rotate, suspect a jammed gear box or open field coil when the correct ac voltage is found at the motor terminals (FIG. 9-22).

When no ac voltage is found at the turntable motor, suspect a defective primary or secondary interlock switch. Check the timer and cook switch terminals for poor contacts. Make sure the timer and cook switch are operating correctly. In combination ovens, check for a defective connection timer assembly. Inspect the wire cables and connections to the turntable motor terminals and all the above components.

The motor armature may not rotate if the bearings are dry or excessively worn. A broken plastic bushing or gear box may jam the motor so that the armature will not turn. You may remove the motor assembly to check the motor and gear box. Remove the four mounting screws to sepa-

Turntable motor

Gear box

Plastic bushing behind gear box

9-21 Intermittent oven operation can be caused by a dry or broken plastic bushing. Also check for a jammed gear box assembly in a Sharp R9600 model.

9-22 Worn or broken gears may cause the gear box to freeze up. Gummed up or old grease may slow down the turntable operation.

rate the motor from the gear box assembly. Now, check the motor for dry or worn bearings. Just lubricating motor bearings may solve the problem of slow rotation. Gummed up motor bearings may cause the motor to run slow and cause the field coils to overheat.

Wash and clean out the bearings when sticking or slow rotation is noted. It's best to remove one end bearing and pull the armature out. Now, wash off or spray cleaning fluid into the metal bearings and armature. Clean off with a rag and cotton swab tips. The gummed up grease may be difficult to remove. After cleanup, lubricate the armature shaft and bearings with a light oil.

A continuity check should be made when correct voltage is found at the motor terminals and the motor will not rotate (FIG. 9-23). Most turntable motors will have a resistance of from 30 to 50 ohms. With a digital vom, 33.5 ohms was measured across a Sharp turntable's motor terminals. Short turns of the field coil may cause a lower resistance measurement. An open field coil will have no resistance reading. If the field coil is open, check for a coil break close to the motor terminals or coil assembly.

9-23 A continuity test should be made when correct ac voltage is found at the motor terminals. With a digital vom, 33.5 ohms was measured across a Sharp turntable motor terminal.

Often an overheated motor will show signs of heat or burned marks on the field coil assembly. In fact, you may find the area burned black from overheating. A motor suspected of running quite warm may slow down and stop. Simply feel the field coil assembly after the motor has operated for at least 30 minutes. A shorted coil may run extremely hot while a normal winding may appear just warm after several hours of operating. Check the coil suspected of overheating with the low-ohm scale of the vom for possible shorted turns. Discard and replace the turntable motor when the resistance reading is under 25 ohms.

The turntable may stop or slow down with a dry or defective bushing bearing. Remove the turntable motor assembly and inspect the bushing. The plastic bearing may be cracked or enlarged causing the turntable assembly to bind or stop rotating. Often a piece of the plastic is broken off. The plastic bushing may be replaced as a separate item. Replace the motor and gear box assembly when a defective motor or broken gears are found. The turntable assembly is a special component and must be ordered from the manufacturer's part source.

In older ovens, the complete oven should be turned on its side for turntable motor removal. Make sure the glass turntable tray is removed so it will not fall out of the oven and break. Remove the metal plate covering the turntable motor. Then disconnect the motor terminals. Remove the bolts securing the motor and gear box assembly to the oven base plate. Pull the motor out to reveal a possible damaged plastic bushing bearing.

The following directions should be followed when removing a defective turntable motor from a Sharp convection microwave oven:

1. Pull the power plug and remove the back cover.
2. Discharge the hv capacitor.
3. Remove the one screw holding the door to the base cabinet.
4. Remove all the screws holding the cabinet to the oven cavity and back area (the back cabinet is largest piece of metal).
5. Release the tab of the back cabinet from the bottom base.
6. Remove the bottom base cabinet.
7. Remove all screws holding the turntable motor to the oven area.
8. Remove the motor assembly and disconnect the leads.
9. Replace the new turntable motor by reversing the procedure.

THE VARI-MOTOR

The vari-motor provides variable cooking in the warm, defrost, simmer, and roast modes. The vari-motor rotates at two revolutions per minute providing intermittent on and off operation of the power transformer and magnetron circuits (FIG. 9-24). For instance, in the defrost operation the oven is on approximately 11 seconds and off 19 seconds, while in the roast mode the oven is on approximately 22 seconds and off 8 seconds. The vari-motor switches in the power transformer and hv circuits for variable cooking and then switches off for so many seconds. The vari-switch is activated by the vari-motor.

Remember the vari-motor rotates very slowly at two revolutions per minute. Check for ac voltage (120V) across the motor terminals. No volt-

9-24 The vari-motor in a Sharp oven rotates at two revolutions per minute, providing on and off operation of the power transformer and magnetron circuit. The vari-switch is activated by the vari-motor.

age here may indicate problems ahead of the vari-motor assembly. If the micro cook light is on, suspect a defective vari-motor or open wiring connections. You may locate a defective vari-switch when the oven seems to be operating, but little heat is produced. Of course, a defective magnetron power transformer, hv rectifier, or capacitor may cause no heat or cooking.

Monitor the ac voltage at the power transformer. The 120 Vac will be present when the vari-switch is in the on position. Notice after a few seconds if the voltage drops off. Now the vari-switch contacts are open. When the ac voltage across the power transformer is off and on for a few seconds you know the vari-motor circuits are operating correctly. If the ac voltage is present and there is no cooking, check the hv circuits.

The vari-motor and switch contacts may be checked with the R×1 scale of the ohmmeter (FIG. 9-25). Always pull the power cord and discharge the hv capacitor before attempting continuity or resistance measurements. The switch contacts will be open in the off position. When in the on position, the vari-switch contacts are closed with a dead short reading. Remove one lead from the vari-motor terminals. Now check for continuity across these two terminals. Replace the vari-motor assembly when you find an open reading.

9-25 The vari-motor switch contacts may be checked with the R×I range of the vom or DMM. When in the "on" position, the vari-switch contacts are closed with a dead short reading. The switch contacts are open in the "off" position.

Since the vari-motor assembly is located on the front panel, the complete control panel must be removed before the screws holding the vari-motor can be removed. Most front control panels are fastened to the metal front panel with metal screws. In some ovens, removing the bottom screws will release the whole front control panel. Disconnect the oven from the power plug and discharge the hv capacitor before attempting to remove any component.

After the front panel is removed, check for the mounting screws holding the vari-motor. Disconnect all wire leads from the motor and switch assembly. Pull off the control knob. Remove the screws holding the vari-motor assembly to the back plate panel. Remove and replace the vari-motor in the reverse procedure.

AUXILIARY BLOWER MOTOR

In many of the convection ovens, another blower motor is turned on to provide cooling air to the microwave components. When the temperature reaches 150° to 160°F the thermal unit turns on the auxiliary motor, which remains on until the oven temperature falls below 125°F, with the thermal unit shutting off the blower motor.

In the Panasonic convection ovens, the heater or convection fan motor moves the air around the heating element and over the food. The lower and upper heating elements are controlled by power relays D and C (FIG. 9-26). The auxiliary blower motor in the Hardwick MM-270 eye-level model is located in the top microwave oven section. Check the convec-

NN-8507/NN-8807/NN-8907 APH

9-26 In the Panasonic convection oven, the heater fan or blower motor is controlled with the DPC at power relay A. Power relays D and C operate the lower and upper heating elements (courtesy of Matsushita Electric Corporation of America).

tion oven blower motors with the R×1 ohmmeter scale. Set the ohmmeter to the R×1 scale to check the thermal unit. This thermal element should read infinite ohms at less than 160°F temperature.

MOTOR PROBLEMS

Here are various motor problems related to the fan, blower, turntable, and timer motor assemblies.

Noisy fan   Check for loose fan blade. A vibration noise may be caused by loose mounting assembly bolts. Inspect for fan blade hitting against fan assembly.

Fan slows down   Suspect dry or gummed up motor bearings when the fan slows down after operating for several minutes. A squeaky motor noise may be due to dry motor bearings. Overloading the hv circuit may cause the fan to slow down.

Intermittent fan operation   Check for a poor wire connection or cable plug to the motor for intermittent fan rotation. A jammed fan blade may produce intermittent fan operation.

Oven shuts down after a few minutes   Suspect an overheated magnetron and open thermal switch if the fan blade is not rotating. Usually the magnetron overheats, opening up the thermal switch and closing down the cooking process. A leaky or shorted magnetron tube may cause the same symptoms.

Intermittent turntable   Check for a broken or cracked plastic coupling between the gear box assembly and the turntable assembly for intermittent or slow turntable rotation. Inspect the motor cable connections for intermittent turntable motor operation. The turntable motor rotation and fan operation may be used as low-voltage circuit indicators.

Longer oven cooking time   If the oven seems to take longer to cook the food, check the cooking time, power transformer, magnetron, and radiating antenna. The radiating antenna or blade may not be moving or moves rather slowly. Although these stirrer motors rotate the blade slower than the blower motor, the radiating antenna blade may be stopped, the belt may be off or broken, or the stirrer motor may be dead. Inspect for a missing cotter key on the antenna wheel and the shaft bearing. Check the stirrer motor for proper rotation and the belt drive assembly on the blower motor for improper deflecting blade operation.

Chapter **10**

Servicing control circuits

The advantage of the control board over manual controls is easy and accurate operation. Simply tap or push the desired buttons or pads of the control panel. The oven will automatically take over all of the cooking process. Most control boards of the various microwave ovens are operated in somewhat the same manner. For correct operation, always check the manufacturer's operating guide or service literature on how the oven operates (FIG. 10-1).

Each oven has its own method of operating and the operational directions should be read. Here is a general operational description of how a Norelco Model MCS 8100 functions:

Cooking with time or temperature Microwave cooking is controlled by either time or temperature. Cook with temperature when internal temperature is the indication required. Use time when visual appearance is the indication required (FIG. 10-2).

Power level Power level gives total flexibility in choosing speed (or power) of cooking that will give best results for each type of food. Power level is used when cooking with time or temperature. It is the name for variable power and includes high, saute, reheat, roast, bake, simmer, braise, defrost, low, and warm. High setting provides the greatest speed; settings between high and warm represent decreasing amounts of microwave power.

Cooking with memories Model MCS 8100 has one memory, which can be programmed with any combination of a variable power setting and time or temperature. The memory automatically changes power setting, cooking time, or temperature. A tone will be audible between memories

I0-I The back view of the control board in a Sharp oven (R5A50 model).

as a reminder that the memory is changing. Colors flash during the memory phase.

Temperature control and automatic hold warm   Temperature control is a feature name for automatic food temperature control and is used as a guide when cooking, reheating, or warming food by temperature. When the temperature control food sensor is in place in the food and the temperature control plug is inserted in the oven receptacle, the oven cooks by judging the internal temperature of the food. After final set temperature is reached, the oven will automatically go into a hold warm setting.

Automatic hold warm   Allows food to be kept warm. Hold warm will only operate if temperature is in the set program. It will continue to operate until food and probe are removed from the oven and the timer has been erased or until 60 minutes have elapsed. Prolonged holding may result in overcooking.

Program check   The oven is designed so that it is possible to check or recall any cooking step programmed before or after cooking has started. Pushing the memory pad will bring set time or temperature of each step into the display; pushing the power level pad will then display the power level setting.

Change or cancel   A cooking step can be changed or cancelled at any time.

Delay cook   The delay cook can be used to postpone the beginning of the cook cycle by entering the delay time desired. It can also be used as a

DISPLAY WINDOW

TIME PAD
Used to
set time.

TEMP PAD
Used to set
temperature

POWER LEVEL
SETTINGS
Select by pushing
numbers with Power
Level word desired.

CHANGE CANCEL PAD
Push once to change
what is in the Display;
push twice to cancel
everything programmed.

POWER LEVEL
PAD
Used to select
or check Power
Level setting.

MEMORY PAD
Used to recall
what has been
programmed.

NUMBER PADS
Push to enter
cooking time or
temperature,
Power Level or
Delay Cook.

KITCHEN TIMER

START PAD
Starts oven.

TIME   POWER LEVEL
TEMP   MEMORY

1 WARM   2 LOW
3 DEFROST   4 BRAISE
5 SIMMER   6 BAKE
7 ROAST   8 REHEAT
9 SAUTÉ   0 DELAY COOK

CHANGE·   CANCEL··

START

10-2 Here is the electronic control panel function of a Norelco MCS 8100 oven. Always know how the various pads are pushed for correct oven temperature (courtesy of Norelco Corp.).

conventional timer (no cooking involved). The oven light will illuminate, but the oven is not operating. The timer is simply counting down.

MCS 8100 BASIC OVEN OPERATION

A. Open door. Oven interior light comes on. Colors appear in the display. Place food in the oven and close the door.

B. Push time pad.

C. Set desired time. Push appropriate number pads. Time will appear in display. Example: one and one-half minutes, push number 1.3 and 10.

D. Push start pad. Blower will start and time will count down. Cooking will be with high or full power. A tone sounds three times when the time is up. Oven shuts off automatically. Colors and oven light will remain on for about one minute.

E. Power Level Check.

    1. Push power level pad. Power level setting will appear in the display.

F. Cancel Time and Power Level.

    1. Push change cancel pad twice.

G. Change Time or Power Level.

    1. Push time pad (or power level pad). Time or power level setting will appear in display.

    2. Push change/cancel pad once. Display will blink for time and show Hi if power level setting.

    3. Enter in new time or power level setting.

    4. Push start pad.

TIME AND POWER LEVEL OPERATION

A. Open door. Oven interior light comes on. Colors appear in display. Place food in oven. Close door.

B. Push time pad.

C. Set desired time. Push appropriate number pads. The time will appear in display. Example: five and one-half minutes, push number pads 5, 3, and 0.

D. Push power level pad and Hi appears in display window. This is a reminder that oven will always operate at full power unless the power level setting is changed.

E. Select desired power level. Push the number pad with selected word. Example: number 5 is simmer. The display will read 50. This means the oven is operating at 50% power.

F. Push start pad. Blower will come on and time will count down. A tone sounds three times when the time is up. The oven shuts off automatically. Colors and oven light will remain on for about one minute after cooking has stopped.

G. Power Level Check.

    1. Push power level pad. Power level setting will appear in display.

    2. Push start pad to return to time counting down.

H. Cancel Time and Power Level.

    1. Push change/cancel pad twice.

I. Change Time or Power Level.

1. Push time pad (or power level pad). Time or power will appear in the display.
2. Push change/cancel pad once. Display will blank for time and show Hi is the power level setting.
3. Enter new time or power level setting.
4. Push start pad.

TEMPERATURE OPERATION

A. Open door. Oven interior light comes on, colors appear in display. Place food in oven; insert probe plug into oven receptacle. F appears in the display. Close the door.

B. Push temp pad.

C. Set desired temperature. Push appropriate number pad. Temperature from 90 °F to 195 °F can be entered. Example: to set 160 °F, push number pads 1, 6, and 0. (If temperature below 90 °F or above 200 °F are entered, a tone will sound and display will go blank except for "F."

D. Push power level pad; Hi appears in display. This is a reminder that the oven will always operate at full power, unless setting is changed.

E. Select desired power level. Push number pad with power level word. Example: roast setting is number 7 pad. Display will read 70. This means oven is operating at 70% power.

F. Push start pad. 90 will appear in display (or actual temperature if greater than 90°), and as food cooks, temperature display will increase in 5 degree increments. When set temperature is reached, a tone will sound three times and oven automatically reduces power to 10% (hold down), display will show time counting down from 60 minutes to indicate how long the oven is in hold warm. Oven will continue to cook in hold warm for 60 minutes or until change/cancel pad is pushed, or door is opened (interrupts but does not erase without change/cancel pad).

G. Temperature or Power Level Check.

   1. Push temp pad (or power level pad). Set temperature or power level setting will appear in display.
   2. Push start pad to return to actual temperature.

H. Cancel Temperature and Power Level.

   1. Push change/cancel pad twice.

I. Change Time or Power Level.

   1. Push temp pad (or power level pad). Set temperature or power level setting will appear in display.

2. Push change/cancel pad once. Display will blink for temperature and show Hi power level setting.

3. Enter in new temperature on power level setting.

4. Push start pad.

MEMORY OPERATION

A. Open door. Oven interior light comes on. Colors appear in display. Place food in oven (if cooking with temperature probe plug it into oven receptacle. F appears in display). Close door.

B. Push time (or temp) pad.

C. Set desired time (or temperature). Push appropriate number pads. The time or temperature will appear in display.

D. Push power level pad. A appears in display. This is a reminder that oven will always operate at full power unless power level setting is changed.

E. Select power level setting. Push number pad next to power level words. Example: simmer setting is number pad 5. The display will read 50 indicating oven is operating at 50% power.

F. To enter next memory, push time or temperature pad. Display blanks, color (or F) will blink, indicating the second winding.

G. Set desired time. Push appropriate number pads. The time or temperature will appear in display (colors or F will continue to blink).

H. Push power level pad. Hi appears in display. This is a reminder that oven will always operate at full power unless power level setting is changed.

I. Push number pad next to power level words selected. Example: roasting is pad number 7. Display will read 70 indicating oven is operating at 70% power.

J. Push start pad. Time for first memory will start, counting down (90 will appear in display for temperature and will increase with actual temperature). At end of first memory, a tone will sound once and oven will automatically shift to next memory. Colors (or F) will blink as a reminder that oven is cooking in the second memory. At end of cooking a tone sounds three times. The oven will automatically stop cooking (if cooking with time) or will reduce to automatic hold warm (if cooking with temperature). Oven will continue to hold warm until the food and probe are removed from oven and the timer has been canceled or until 60 minutes have elapsed.

K. Check Set Time or Temperature.

1. Push memory pad. Set time or set temperature will be displayed.

2. Push memory pad again, next memory will be displayed.

3. Push start pad to return to time counting or actual temperature.

L. Check Power Level Setting.

1. Push memory pad until memory to be checked is in display.

2. Push power level pad.

3. Push start pad to return to time counting or actual temperature.

M. Cancel All Time, Temperature, or Power Level.

1. Push change/cancel pad twice.

N. Change Set Time or Temperature.

1. Push memory pad until memory to be changed is in display.

2. Push change/cancel pad once.

3. Enter new time or temperature by touching number pads.

4. Push start.

O. Change power level while cooking.

1. Push memory pad until memory to change is in display.

2. Push power level pad.

3. Push number pad for new power level setting.

4. Push start.

THE CONTROL BOARD OR DPC

The basic control board may have a keyboard, key harness or membrane, external small ac voltage power transformer, and control circuit (FIG. 10-3). Most control circuits are powered by a small ac power transformer. A temperature-probe thermistor operates from the control circuits. The control board operates a power cooking relay or triac assembly to control the oven cooking.

In the Panasonic ovens, the digital programmer (DPC) may have a microcomputer circuit with a pulse oscillator, buzzer, display, humidity sensor, temperature control, clock pulse, initializer, door check, variable power relay control, power relay control, membrane switch keyboard, and low-voltage power supply (FIG. 10-4). The different functions of each operation are shown in FIG. 10-5.

10-3 Here is a block diagram showing what components the control board operates in a basic oven.

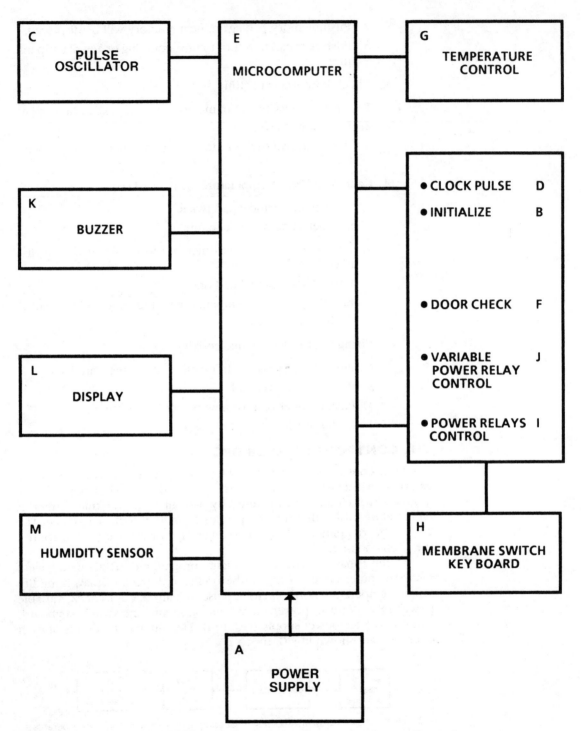

I0-4 The block diagram of the DPC in the Panasonic ovens (courtesy of Matsushita Electric Corporation of America).

	BLOCK	FUNCTIONS
A	POWER SUPPLY CIRCUIT	RECTIFIES THE A.C. VOLTAGES SUPPLIED FROM THE LOW VOLTAGE TRANSFORMER, REGULATES THE DC VOLTAGES AND SUPPLIES THEM TO THE RESPECTIVE CIRCUITS
B	INITIALIZATION CIRCUIT	RESETS THE MICROCOMPUTER WHEN THE OVEN IS FIRST PLUGGED IN
C	PULSE OSCILLATOR	OSCILLATES AT A SPECIFIC FREQUENCY TO CONTROL THE PROCESSING SPEED OF THE MICROCOMPUTER
D	CLOCK PULSE GENERATION CIRCUIT	GENERATES A CLOCK PULSE AT THE POWER SUPPLY A.C. FREQUENCY FOR SYNCHRONIZING THE CLOCK TIMING
E	MICRO COMPUTER	DETERMINES AND PROCESSES VARIOUS INFORMATION AND GENERATES THE APPROPRIATE OUTPUT SIGNALS IN THE SEQUENCE THAT WAS PREPROGRAMMED
F	DOOR SIGNAL CHECK CIRCUIT	INPUTS A SIGNAL TO THE MICROCOMPUTER TO INDICATE WHETHER THE DOOR IS OPEN OR CLOSED BY DETECTING THE SIGNAL FROM THE DOOR SIGNAL SWITCH
G	OVEN TEMPERATURE CONTROL CIRCUIT	READS CURRENT TEMPERATURE WITHIN THE OVEN CAVITY AND FEEDS THIS INFORMATION TO THE MICROCOMPUTER
H	MEMBRANE KEY BOARD	INPUTS VARIOUS DATA AND INSTRUCTIONS INTO THE MICROCOMPUTER THROUGH THE MECHANICAL CONTACTS OF THE MEMBRANE SWITCH KEY BOARD
I	POWER RELAY CONTROL CIRCUIT	CONTROLS ON-OFF TIMING OF POWER RELAY TO TURN THE MICROWAVE OVEN ON-OFF WITH SIGNALS FROM THE MICROCOMPUTER (MAGNATRON CONTROL)
J	VARIABLE POWER RELAY CONTROL CIRCUIT	CONTROLS ON-OFF TIMING OF VARIBLE POWER RELAY WITHIN A SPECIFIC DUTY CYCLE TO VARY THE OUTPUT POWER OF THE MICROWAVE OVEN WITH SIGNALS FROM THE MICROCOMPUTER
K	BUZZER CONTROL CIRCUIT	GENERATES BEEP TONES INDICATING THE COMPLETION OF A PROGRAMMED OPERATIONS IN THE MICROWAVE OVEN
L	DISPLAY CIRCUIT	TURNS ON NECESSARY SEGMENTS OF THE DISPLAY TUBE TO INDICATE THE APPROPRIATE FIGURES, SYMBOLS, LETTERS ETC.
M	HUMIDITY SENSOR CONTROL CIRCUIT	READS CURRENT HUMIDITY CONDITIONS WITHIN THE OVEN CAVITY AND FEEDS THIS INFORMATION INTO THE MICROCOMPUTER

10-5 The various block diagram components and their functions in the Panasonic DPCs (courtesy of Matsushita Electric Corporation of America).

Internal control board problems may result from improper operation, improper temperature control, erratic programming, improper sequence operation, improper display numbers, and no cooking or improper cooking. Always double check for proper operation of the control board. Refer to the manufacturer's service literature.

When an oven with a touch-control panel malfunctions, try operating the controls. If you press the proper sequence of keys or buttons and do not hear audible signals, suspect a defective keyboard. Check for proper illumination of the lighted display. Are there two or more figures on the display at the same time? If so, the control panel is defective.

Check the power transformer input voltage when you have no display lights or the control board appears dead. Measure the output voltage from the controller when the display and board seem normal (FIG. 10-6). Improper input and output voltages from the controller may indicate external or internal control problems.

CONTROL BOARD CIRCUITS

Before replacing the suspected control board, check all input voltages. The low ac voltage from the small power transformer may be missing causing improper control board functioning. This low-voltage power transformer is fed directly from the power line. You may find this transformer on the control board of some ovens (FIG. 10-7). Check the low ac voltage right at the terminals where voltages tie into the control board. Although a few manufacturers provide these transformer voltages, it is best to mark them on the schematic for future reference (FIG. 10-8).

Panasonic's digital programmer power supply A small low-voltage power transformer (T1) supplied ac voltage to the heaters of the display tube, ac voltage to the clock pulse generator, dc voltage to the power relay circuit, dc voltage to the microprocessor, and dc voltage to the buzzer display and display segment (FIG. 10-9).

The secondary winding of T1 supplies a 2.7 Vac to the heaters (F) of the display tube (S5 and S6). At the secondary winding S4 (terminal D) 24 Vac is supplied directly to the clock pulse generator. The 24 Vac winding (S4) is rectified by diode D1 and supplies an 18 Vdc source for the power relay circuits. The regulation 5 Vdc source at pin 64 provides supply voltage to the microprocessor. A −25 volt supply for the buzzer display and display segment is rectified by diode D2.

Check the power transformer winding with the ohmmeter when low or no ac voltage is found at the small transformer. Always discharge the high-voltage (hv) capacitor and unplug the power cord when taking ohmmeter measurements in the oven. Remove one ac input lead from the power transformer and take a resistance reading across the primary winding. Replace the transformer when you find an open reading. Sometimes these primary windings open after lightning, power outage damage, or an overloading of the control board circuits (FIG. 10-10). Low continuity resistance of the secondary winding may indicate the windings are normal.

10-6 The power transformer which provides dc voltage for the function of the control board. Check the power-transformer voltages when you have no display lights or with a dead display board. The power-transformer might be mounted separately from the control board, as in the Quasar model.

10-7 You may find the small transformer operates directly from the power line and, in some ovens, is mounted on the control board assembly. The transformer is found on the control board of this Sharp oven.

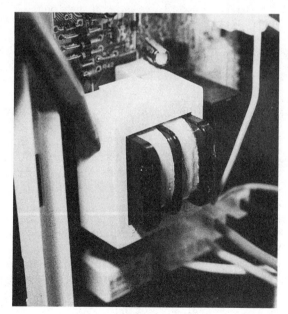

TEST SET-UPS	TEST POINTS	NORMAL VOLTAGE (APPROXIMATE)
Attach meter leads to wire harness test points A-B, apply power to oven and open oven door.	A-B	120 VAC
Disconnect power and remove low voltage transformer connector attached to control circuit board and attach meter leads into harness side of connector at test points shown in chart. Apply power to oven and open oven door.	2(blue)-3(yellow)	2.5 VAC
	1(red)-6(brown)	20 VAC
	1(red)-4(orange)	20 VAC

10-8 Since oven manufacturers do not give small transformer voltages, take the voltage with a normal oven and write them on the service schematic. Here is a test chart of transformer voltages in a Norelco MCS 8100 model (courtesy of Norelco Corp.).

The ac primary voltage for the small power transformer operates directly from the power line. When the oven is plugged in, the color numbers should appear in the digital display in most ovens. If the ac voltages are normal from the transformer and the display does not light up, suspect poor board connections. Measure the transformer ac voltage right at the control board terminals. Poor connections may cause erratic oven operation.

Dirty or corroded touch panel contacts may cause dead or erratic oven operations. Some ovens have a flat type ribbon cable from the touch panel to the control board. Be careful with these ribbon cables. Overflexing or scratching of the ribbon cable may damage the silver circuit wiring. In a Norelco MCS 8100 oven, the ribbon cable may be checked with an ohmmeter by disconnecting the cable from the circuit board. Apply even pressure on both sides of the ribbon cable and pull outward. Take ohmmeter readings of the various test points and compare them (FIG. 10-11). Replace the defective touch panel assembly if improper indications are given.

10-9 The power supply for the digital programmer microprocessor of Panasonic's control board with the various ac and dc voltage sources (courtesy of Matsushita Electric Corporation of America).

10-10   Take a continuity measurement of the primary winding of the small transformer. Sometimes lightning or a power outage can open up the primary winding.

PANASONIC KEYBOARD CHECKS

Check continuity between switch terminals of the membrane switch assembly by tapping an appropriate pad on the keyboard. The contact assignment of the respective pads on the different model keyboards are shown in FIG. 10-12.

TOUCH PAD MEASURED	RIBBON CABLE TEST POINT	NORMAL READING (TOUCH PAD PUSHED IN)	NORMAL READING (TOUCH PAD NOT PUSHED IN)
Time Temp	4-7	Less than 100 ohms	Open
2	3-6	Less than 100 ohms	Open
5	2-8	Less than 100 ohms	Open
0	10-9	Less than 100 ohms	Open

10-11   The low ohmmeter reading of the ribbon cable may determine if the cable or touch panel controls are functioning. A test chart is given for a Norelco Model MCS 8100 oven.

NN-9807

NN-9507

MQ8897/8877

*MQ8897 only

10-12 Check the continuity between switch terminals at the terminal connection by pressing the appropriate key (courtesy of Matsushita Electric Corporation of America).

Erratic oven operation and improper sign display may be caused by a corroded or dirty ribbon cable and pin connector. Simply cut off the corroded end of the cable, which will start a new set of contacts. Check to make sure this type of contact and ribbon cable is found before cutting off the cable.

PANASONIC HUMIDITY CONTROL AND SENSOR CIRCUITS

In the Panasonic ovens, the humidity and sensor control circuits are tied to the microprocessor circuits. In the humidity sensor heater control circuits, when E36 goes low, Q5 and Q6 turn "on" and bias the triac on passing ac to the heater. The triac stays on for three to five seconds (FIG. 10-13). When E36 goes high, Q5 and Q6 turn "off," biasing the triac "off," which shuts off the heater.

In the humidity sensor control circuits, E32 supplies a varying pulse width (pulse width modulation) through Q2 and C5 (blocks dc) to the humidity sensor from the microprocessor. The output signal from the humidity sensor varies the pulse width amplitude depending on the

10-13 The humidity sensor heater and humidity sensor control circuits in the Panasonic ovens (courtesy of Matsushita Electric Corporation of America).

humidity in the cavity. This signal is fed back to the microprocessor unit via AN7 (refer to FIG. 10-13).

PANASONIC NN-9807/MQ8897
HUMIDITY SENSOR AND DPC TESTS

Do not touch any parts of the circuitry on the DPC because static electric discharge may damage this control panel. Always ground yourself while working on the panel to discharge any static charge built up on your body. Use a wrist strap grounding device.

A. Check across the sensor heater terminals. Normal cold resistance should read approximately 6 ohms.

B. In order to determine if the auto sensor/easy-matic cooking function of the DPC is in working order, do the following tests:

1. Place a water load in the oven.

2. Unsolder the red and white wires connected to sensor terminals (FIG. 10-14).

10-14 The oven humidity hookup in the Panasonic NN-9807/MQ8897 models (courtesy of Matsushita Electric Corporation of America).

3. Tap auto sensor cook control pad once for set auto sensor cooking at A1 and tap start pad (NN-9807).

4. About 10 seconds after start pad is tapped, short red and white wires for five seconds and then remove them. Before shorting the red and white wires, you can watch the sensor heater glow red.

5. The oven lamp will dim a little, about 50 seconds after the start pad is tapped. This indicates that the magnetron has started oscillation.

6. Ten seconds after the magnetron oscillation, short the soldered wires again. The oven should turn off automatically
when the red and white wires are shorted.

The auto sensor/easy-matic cooking function in the DPCs is in working order if the above condition is obtained.

PANASONIC OVEN TEMPERATURE SENSOR THERMISTOR TESTS

This sensor monitors the heat produced by the heater circuit and main-
tains the oven temperature the user had selected. Normal room temperature, especially in a kitchen, can vary anywhere from 50° to 90°F. The
reading across the oven sensor thermistor should be between 360k ohm
and 800k ohm when read in an area within 50° to 90°F room temperature
range. If the resistance reading is out of the range slotted here, the sensor
is defective and must be replaced.

When measuring resistance, disconnect the CN3 connector from the
DPC, otherwise a false reading may be indicated. If checking an oven sensor thermistor just after the microwave oven has been operating, the sensor of course won't be room temperature. In this case, the sensor must be
removed and allowed to cool down to the 50° to 90°F range.

PANEL—No cook operation

After the cooking time is tapped or pressed on the control board, the
oven circuit should respond with magnetron operation. Check the control board output voltage or external components when the control
board seems to operate normally with no cooking or heat from the magnetron. Usually, the control board provides voltage to an oven relay or
triac assembly (FIG. 10-15).

When the power relay or triac is energized by the DPC, power-line
voltage (120 Vac) is applied to the fan motor and power transformer. You
may notice the oven lights pull down or start to blink when the magnetron circuits are functioning. The cooking time starts to count down.
When the cooking time counts down with no heat, suspect a defective
control board or oven relay and triac assembly.

Measure the voltage across the relay solenoid coil terminals (FIG.
10-16). No voltage at the power relay coil indicates a defective control
board. If the relay contacts are not closed, suspect a defective relay when
voltage is found at the coil terminals. Check for an open coil with the low-
ohm scale of the ohmmeter.

PANASONIC'S POWER CONTROL CIRCUITS

In the power relay control circuits, when pin 13 of IC2 goes high, and the
power relay control transistor turns on, its collector goes low energizing
relay RY1, which turns on the oven lamp (FIG. 10-17).

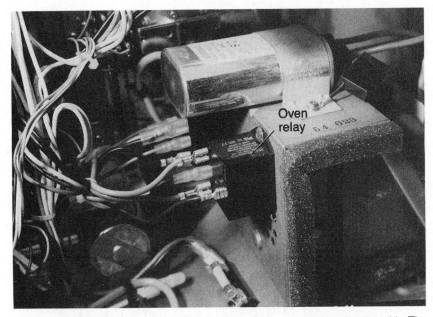

10-15 Usually, the control board provides voltage to an oven relay triac assembly. The small oven relay completes the power-line circuit to the primary winding of the hv circuit to the primary winding of the hv power transformer.

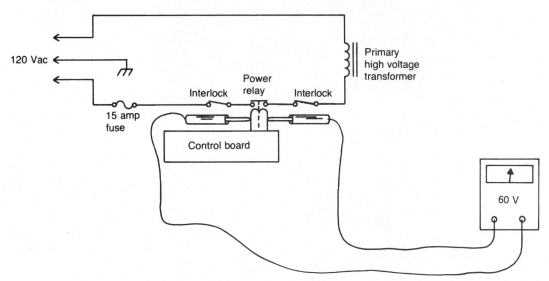

10-16 Measure the voltage across the solenoid relay coil terminals. Low or no voltage could indicate a defective control board or transformer. When the voltage is measured and there is no relay action, take resistance measurements of the solenoid winding.

10-17 The power control in the Panasonic microprocessor circuit (courtesy of Matsushita Electric Corporation of America).

Within the magnetron relay control circuit, an initialization E25 of the microprocessor goes low. Upon pressing the start key, pin 6 will go low and the latch circuit output will go high, turning on the power relay drive transistor. At the same time, R11 of the microprocessor unit will go high turning on Q4 and energizing the magnetron via relay RY B.

In the initialized circuit, the output pin 14 of IC2 goes to the microprocessor unit reset. Under normal conditions, pin 14 goes high. If power is interrupted, pin 14 to the microprocessor unit reset goes low and the microprocessor unit will lock up.

With the clock pulse generator circuit, pin 16 of IC2 receives 18 Vac from the secondary winding of the low-voltage transformer. The inverter clips and inverts the input ac and generates a 7V P-P output square wave to IR Q of the microprocessor.

PANASONIC'S OTHER CIRCUITS

There are many other circuits that the microprocessor controls within the Panasonic ovens.

Pulse oscillator circuit

The pulse oscillator circuit functions from pins 25 and 26 of the microprocessor (FIG. 10-18). The pulse oscillator circuit is 4.19 MHz. The oscillator's peak-to-peak voltage is 5 volts. This signal controls the processing speed of the microprocessor unit.

10-18 The Panasonic pulse-oscillator circuit operates on pins 25 and 26 of the microprocessor. This 4.19 MHz signal controls the processing speed of the microprocessor unit (courtesy of Matsushita Electric Corporation of America).

Buzzer control

When the appropriate membrane key is pressed and accepted by the microprocessor, the microprocessor unit generates a single burst of 2.1kHz pulse at E24, which activates the buzzer circuit (FIG. 10-19). When the cooking program is completed, the microprocessor unit will generate a series of five 2.1kHz bursts, creating five audible beeps.

10-19 The piezo (2.1kHz) buzzer indicates when the oven cooking process is completed in the Panasonic ovens with audible beeps (courtesy of Matsushita Electric Corporation of America).

Display tube

The EO to E19 outputs of the microprocessor controls the operation of the display tube (FIG. 10-20). EO through E7 and E19 control the grids of the display tube. E12 through E18 controls the numerical segment sections (A to G) of the display tube.

Temperature control

The oven temperature sensor monitors the temperature changes in the oven cavity. This sensing device is called a *thermistor* and its resistance change is converted into a voltage variation across the resistor network R14 and R15 (FIG. 10-21). This voltage change is fed back to the micropro-

- THE E0 TO E19 OUTPUTS OF THE MICROPROCESSOR CONTROL THE OPERATION OF THE DISPLAY TUBE.

10-20 The display tube is controlled by the outputs from E0 to E19. Notice the two different −25 V supplies from the low-voltage power circuits supply operation voltage (courtesy of Matsushita Electric Corporation of America).

10-21 Check the temperature thermistor on the high ohmmeter scale. The resistance across the thermistor should be around 380k ohms at 50°F and 800k ohms at 90°F in the Panasonic oven (courtesy of Matsushita Electric Corporation of America).

cessor unit port AN6 to control the cycle on time of the magnetron or the heaters.

The temperature control sensor monitors the heat produced by the heater and establishes the oven temperature the user has selected. In a kitchen, normal room temperature can vary anywhere from approximately 50° to 90°F. The resistance reading across the oven thermistor should be within 380k ohms at 50°F and 800k ohms at 90°F.

Bake-broil control

The solenoid operation, when operating the convection mode, controls a vent door in the cavity to prevent heat loss. The solenoid is activated when R12 is high and IC3 pin 11 goes low (FIG. 10-22).

In standby mode, E37 and R12 are low and R13-14-15 are high. When the broil mode is selected E37 goes high, Q7 and Q8 then turn on, +18 Vdc is applied to the relay RY 3 (upper heater relay PY C). Simultaneously, the microprocessor R13 goes low and this causes a low at pin 13 of IC3, which turns on the upper relay via relay RY 3. Only the top heating element is used in the broil mode.

10-22   The bake-broil control circuits of the DPC in the Panasonic ovens (courtesy of Matsushita Electric Corporation of America).

In the bake mode, the upper and lower heaters alternately go on and off in accordance with the cooking temperature. When the bake mode is selected, E37 goes high, Q7 and Q8 turns on, +18 Vdc is applied simultaneously to the coils of the power relays RY C and RY D.

The microprocessor R13 goes low, generating a low pin 13 or IC3, turning on the upper heater relay RY C. Alternately, R14 of the microprocessor goes low, causing RY D to turn on the lower cavity heater, while R13 goes high.

PANASONIC'S VARIABLE POWER CIRCUIT

The variable power circuit controls on-off timing of the variable power relay within a specific duty cycle to vary the output power of the microwave oven with signals from the microcomputer. The variable power circuit basically incorporates a phase control circuit, a multivibrator circuit, and a switching circuit. In some advanced models, these circuits are combined into one piece of microprocessor (LS1). However, we will study the analog circuit, commonly used as the variable power circuit, in this section (FIG. 10-23).

PHASE CONTROL CIRCUIT

This circuit functions to control the open/close timing of the variable power switch to prevent possible damage to the switching device and possible interference to TV and radio as the result of arcing generated across the contacts of the switching device.

In the hv circuit, the high voltage is applied to the magnetron only during the negative half cycle, and no high voltage is developed across the variable power switch during the positive half cycle, as explained in the hv circuit in this chapter. The basic idea of the phase control circuit is to

10-23　Here are Panasonic's variable power control circuits, including phase control, multivibrator circuit and switching circuits (courtesy of Matsushita Electric Corporation of America).

open or close the high voltage circuit when no voltage is applied to the magnetron. We will see how this is done in the following section.

1. When the start button is pressed with door being closed and timer set, approximately 18 volts is generated in the tertiary winding of hv transformer and supplied to the variable power circuit.

2. The 18 Vac is converted to 24 volts half wave by the diode D2 and this half cycle voltage is divided by the R3 (100k)-R4 (100k) combination and resulting 12 volts half wave is applied to the gate of transistor Q1.

3. Capacitor C1 is charged through R1 and when the voltage across the C1, which is equivalent to anode voltage of Q1, exceeds the gate voltage of Q1, it causes Q1 to conduct.

4. When Q1 conducts, a voltage is developed across R2 and will forward-bias the base-emitter junction of transistor Q2 allowing it to conduct. **NOTE:** Conduction of Q1 discharges C1 and this will cause Q1 to cut off when C1's voltage drops to approximately 0 volts. The Q1 remains cut off until next positive half cycle charges C1 to 4.2 volts.

5. When Q2 conducts, it causes the gate voltages of Q3 and Q4 to drop from 14 volts to 12 volts through R7 and R16. **NOTE:** The transistor Q2 conducts only when Q1 conducts, which means that Q2 conducts only during positive half cycle and does not function when the high voltage is across the variable power switch.

6. As long as Q2 remains cut off, the gate voltages of Q3 and Q4 remain 14 volts and this will result in no conduction of both Q3 and Q4 as the anode voltages of both transistors never exceed 14 volts.

In this way, the variable power switch is controlled to open or close only when no high voltage is across the variable power switch contacts.

MULTIVIBRATOR CIRCUIT AND SWITCHING CIRCUIT

This circuit functions to control the open/close time duration period of the variable power switch. As shown in FIG. 10-24, the programmable unijunction transistors Q3 and Q4 are connected in parallel with the voltage divider resistors R8-R9 for Q3 and R14-R15 for Q4. We will see how this circuit functions in the following section.

1. 24 Vdc generated by diode D1 is smoothed by capacitors C3 and C4 and is then applied to the multivibrator circuit. The gate voltages of Q3 and Q4 are determined by the RY-R8-R9 combination for Q3 and R6-R14-R15 combination, plus the switching effect of Q2 on both combinations. When the circuit is first energized, the transistor with the higher anode potential conducts first.

2. Assuming that the anode voltage of Q3 is higher than Q4, when

10-24 The multi-vibrator circuit and switching circuits in the Panasonic variable control circuits (courtesy of Matsushita Electric Corporation of America).

the circuit is first energized Q3 conducts and 12 Vdc (as the result of Q2 conduction) on the gate of Q3 becomes approximately 0 volts. This causes current to flow as shown by the solid line and causes capacitor C5 to charge.

3. When the anode voltage of Q4 exceeds the gate voltage as the result of charging effect of C5, it causes Q4 to conduct and the current flows as shown by the dashed line and this causes Q3 to cut off. The charging time of C5 is determined by the time constant of the series combination of the variable resistor (portion A) and R12-C5-R11.

4. The voltage across R13 forward-biases the base-emitter junction of Q5 causing it to conduct. The conduction of Q5 instantaneously causes Q6 to conduct allowing 24 Vdc to be applied to the variable power switch coil thereby closing the hv circuit to start microwave oscillation.

5. As soon as Q4 conducts, the capacitor C5 starts to charge in the opposite direction through R10 and R11. When the anode voltage of Q3 exceeds its gate voltage (12 volts), Q3 conducts and Q4 cuts off. The complete time cycle of one transistor conduction, through cutoff and back into conduction, is approximately 22 seconds. This on/off cycle of Q3 and Q4 is repeated as long as the circuit is in operation.

The length of time within the 22 second time period that Q4 conducts is determined by the setting of the variable power resistor mounted on the front panel of each microwave oven. When the variable power resistor is set at medium-low power, the multivibrator circuit is balanced and each transistor will have equal conduction time periods of 11 seconds each, resulting in 50% of full output power to be obtained.

As the variable power resistor moves toward the B portion end (low power), Q4 conducts for a shorter period; when moved toward A portion, it conducts longer. In this way, the output power of the microwave oven can be varied to any power setting desired by adjusting the external variable power resistor controller. When the variable power switch is moved toward the extreme portion A end, the high-power switch closes. This will keep Q5 and Q6 conducting constantly to maintain full power oscillation. **NOTE:** It is possible to check the time duration of one complete cycle of the variable power circuit. To check the on/off cycles of the magnetron, simply observe dimming of the oven light. If the oven light dims, the secondary circuit is supplying high voltage to the magnetron and when the oven light goes bright, it indicates no voltage supply to the magnetron.

As explained above, one complete cycle is 22 seconds. However, it can vary approximately 7 seconds due to the tolerances of resistors and capacitor C5. This tolerance of duty cycle will not affect the actual output power because the proportion of on/off time period remains the same. For example, if the duty cycle is measured as 26 seconds, the duty on time of medium-low power setting should be 13 seconds resulting in 50% of full power output.

TROUBLESHOOTING THE REST

To determine if the rest of the oven circuits are operating, clip a wire across the power relay terminals (FIG. 10-25). Always pull the power plug and discharge the hv capacitor before attempting to connect the clip wire. Now, plug the oven in and if the hv and magnetron circuits are normal, the oven will begin to cook without tapping in any time on the control board. Replace the control board when no voltage or low voltage is found at the oven relay and the rest of the oven circuits are normal.

10-25 To determine if the oven circuits are operating, clip a wire with alligator clips across the relay terminals. Be sure the capacitor is discharged and the power cord pulled before applying the clip lead.

In ovens using a triac assembly controlled by the control board, check the ac voltage applied to the gate terminal of the triac. No voltage here indicates a defective control-board assembly with proper ac input voltage. To test the rest of the hv and magnetron circuits, clip a wire across MT1 and MT2 of the triac assembly (FIG. 10-26). Again, discharge

10-26 To determine if other circuits are working with a suspected triac, clip a wire lead across MT1 and MT2 of the triac assembly. Again, discharge the capacitor and pull the power plug before applying the clip lead.

the hv capacitor and pull the power cord. When the oven begins to cook, you know the rest of the circuits are normal.

Replace the defective control board when abnormal voltages are found at the power relay or triac assembly with proper low ac input voltages. When the digital display does not come on with proper input ac voltage, replace the control board. After programming the control board, at least three times according to the manufacturer's directions, and the oven does not start to count down, suspect a defective control board. Make sure the temperature probe is out of the socket when making these tests. If the display numbers do not register properly after several attempts, suspect a defective controller assembly (TABLE 10-1).

LIGHTNING DAMAGE

Often, the control board is damaged when the oven is struck by lightning or by a storm-caused power outage. Before replacing the control board, check for a burned or damaged varistor on or near the control board. In some ovens, the line-voltage varistor is outside the control board. Check the varistor for burned marks (FIG. 10-27). Replacing the varistor may solve the control board problem. These small varistors are added across the ac line to protect the oven or control board circuit.

Remove the control board and check for damaged pc wiring. Some ovens have a test cord kit to connect the board to the oven for easy repairs. Sometimes the wiring may be damaged around the defective varistor (FIG. 10-28). Bridging the burned pc wiring with regular hookup wire may solve the dead oven symptom. If the control board is excessively damaged or appears erratic after being struck by lightning, replace the entire control board.

Check the control board contacts and plug-in connections. Arcing around the contact pins will show melted areas around the plastic. Inspect both male and female connections. If dark or arcing marks are found around the pin connections, remove the board and inspect for badly soldered pins. Clean up the bad contact connections. When these

**Table 10-1   A control board troubleshooting chart.
If the oven service manual is handy,
follow manufacturer's troubleshooting procedures.**

Symptoms	*Possible Problems*
No display light or flashing	1. Check fuse or power line. 2. Check low ac voltage from small power transformer to controller. 3. Make sure temperature probe is out of socket. 4. Suspect controller board when low voltage is found at control board terminals. 5. Check circuit breaker. 6. Check defective varistor.
Display light with flashing	1. Low ac voltage. Transformer normal and applied voltage to control board. 2. When time will not register, suspect control board. 3. Go over time and temperature setting at least three times before replacing control board.
Improper number display and incomplete segments	1. Double check the oven operation from manufacturer's service manual. Make sure you are tapping proper numbers in sequence. 2. Check each pad membrane keyboard continuity. 3. Defective control board. 4. When some of the numbers light up and not others, suspect defective control board. 5. Defective touch pad.
Improper count down	1. To check the timer functions, touch pads and set desired time. 2. The display should start to count down. 3. Replace board when display shows correct time and no count down. 4. Replace control board if display jumps or too fast count down.
Time and temperature set. Time counting down. No blower motor or cooking operation	1. Suspect no control voltage or defective cook relay or triac. 2. Measure voltage to controlling component. If voltage is present check defective relay or triac.

Table 10-1   Continued.

Symptoms	*Possible Problems*
	3. Use test clip method to isolate defective relay or triac.
	4. Replace control board with no control voltage.
	5. Check door interlock.
	6. Check microprocessor in control board.
To check power display and stop operation	1. Set time and temperature. Push switch.
	2. Oven begins to cook. Then stops switch or off button.
	3. Oven should stop. If not check triac assembly.
Time and temperature pushed. Time counting down. Blower motor operating with no heat.	1. Check 120 Vac at the primary winding of the high voltage transformer.
	2. If power-line voltage present, problems within high-voltage circuits.
	3. No power line voltage at the primary winding of the high-voltage suspect defective oven relay, triac or control board.
	4. Check safety concealed switches.
A faint display number is seen in background of display	1. Although the oven count down and cook properly, sometimes you can see a faint number in the background at all times.
	2. Replace defective control board. This may occur after power outage or lightning damage.
Control panel defective a few months after replacement	1. Check for improper fusebox ground.
	2. Intermittent gate terminal or triac.
	3. Magnetron tube.
Temperature probe will not key in . . no indicator light.	1. Check continuity of probe.
	2. Defective sensor probe.
	3. Check jack terminal leads.
	4. Defective control board or microprocessor.
Oven stops . . soon as start button pushed in with temperature probe in operation.	1. Check continuity of probe.
	2. Defective sensor probe.
	3. Check continuity of probe jack and wires.
	4. Check for shorted probe.
	5. Defective control board and microprocessor.

10-27 The ac varistor was damaged in a Sharp oven control board assembly. Notice the burned marks on the ac cable plug connections.

connections are very bad, try and replace them. For temporary repairs, solder a flexible lead wire around the two connections.

BEFORE YOU REPLACE THE CONTROL PANEL

The following is a list of things to check before you replace this panel in a MCS 8100 oven. It will also help you to service any other panels in the future. Do not assume the board is bad on every call you get.

10-28 Check the backside pc wiring for possible pc wiring damage. Especially check the wiring where the ac enters the plug connections.

1. Check the customer's electrical system for correct polarity and especially for a good ground. Without a ground, the system will act erratic, skip numbers, give incorrect readouts, and be hard to program. Check line voltage. It must be at least 110 volts for proper operation. Always check around connections from board all the way to the round pin plug. You should have continuity.

2. Recheck the complaint. Remember some problems can be caused by the customer not really knowing how to operate the unit. A prime example is if the customer waits too long between touching the clock set pad and number, the unit will not enter a number. There is a three second time limit. Become completely familiar with the operation of the touch panel. Always have the customer operate the unit for you. Use check procedure in service manual.

3. Make sure all door interlocks are operating properly. An ohmmeter check of these interlocks should show (with door open) interlock monitor closed and primary interlock open, secondary interlock open, door interlock sense switch closed. With door closed, all switches will check opposite of above. Remember if door interlock sense switch is not working, you will be able to program unit, but the unit will not start when start is touched. An improperly adjusted door interlock sense switch will cause the symptom of not shutting count-down off when door is opened.

4. Check the on/off sense switch, which is a contact between pin 7 and 8 of on/off switch. An open on/off switch will not allow the light and fan and stirrer to come on. The only thing that will happen is you will get all 8888s on display.

5. If you have a unit that programs, but the light, fan, and stirrer will not work, browner will not heat, and oven will not operate, check pin 3 to 16 on on/off switch. With switch on, these contacts will be closed.

6. Many times a control panel is changed because the unit starts operating as soon as the on/off switch is pushed. This could very well be a bad triac. To check, pull the wire off the gate terminal of the triac in question. If the unit still operates as soon as the on/off is pushed, you have a shorted triac. If it does not, you either have a bad board or bad disconnect block. If you notice a burned spot on the board edge connector, the burned spot indicates that you have had a loose terminal—order part number 355T498S02.

7. If unit shorts as soon as on/off switch is turned on, this problem could be caused by a short in either the magnetron fan or stirrer motors. To isolate which one it is, make a resistance check of these motors. With motor disconnected, read the following: stirrer motor 2,000 ohms, magnetron fan motor about 25 ohms.

8. Always check to see if printed circuit board is loose in the frame. You should not be able to move the board up and down at all. If

board moves, bend up the tabs on bracket, which the board sets on, until the board no longer moves.

iNSTALLING A NEW CONTROL BOARD

Be careful when removing the control panel from the packed carton. Most control boards are enclosed in static-type bags. Always hold the control board by the edges. Keep fingers and metal material away from the printed wiring and board components (FIG. 10-29). First, touch the metal oven cabinet to discharge any static electricity before mounting the control board.

In some ovens, the control board is easily removed from the backside while in others the front control panel assembly must be removed. Install the new board in reverse order. Make sure all board plugs are tight and snug. In ovens with flexible rubber cable, make sure the cable is pushed in tight. Inspect the plug as you plug it in. Tighten all metal screws holding the board to the metal frame.

Before replacing any control board or panel, make sure it is defective. Run through the operation procedure several times. Take input and output voltage measurements of the control board. Determine if the control board or other circuits are at fault. Be careful in handling and mounting the control board. Control boards are fairly expensive and take several minutes to install. If in doubt, follow the manufacturer's service manual for control board troubleshooting procedures.

10-29 Handle the new control with fingers at the board edges. Wear a wrist ground strap. Keep body at the same potential as the metal oven base.

Chapter 11

Microwave leakage tests

You may be called upon to just check the oven for rf leakage if a customer is afraid the oven has possible radiation. Most people are deathly afraid of becoming burned or receiving rf radiation. Always check the microwave oven for energy leakage after each repair. You are required to take radiation leakage tests after each oven repair if the oven is in the warranty period. Be safe—take a radiation test to prevent possible damage or even a preventable lawsuit (FIG. 11-1).

The United States government leakage standard is 5mW/cm², while in the customer's home. The power density of the microwave radiation emitted by a microwave oven should not exceed 1mW/cm² at any point 5cm or more from the external surface of the oven, measured prior to acquisition by the purchaser. Throughout the useful life of the oven at any point, 5cm or more from the external surface of the oven should not exceed 5mW/cm². Should the leakage be more than 2mW/cm², some microwave manufacturers require that they be notified at once. The defective oven should not be used until the radiation leakage has been corrected according to the manufacturer's instructions (FIG. 11-2).

IMPORTANT SAFETY PRECAUTIONS

Before the oven is returned to the customer and when installed in the home, check off the following safety features:

- Do not operate the oven on a two-wire extension cord. Always operate the oven from a properly grounded outlet.
- The oven should never be operated with a defective door seal or warped door. Check the choke and gasket area. Keep the door area

II-I The different areas of the oven to be checked for possible microwave radiation. Be careful when making tests around hv components.

II-2 After each repair, check the oven for leakage with a recommended leakage survey monitor. One of these is a Simpson Model 380M, shown here. The white cone tip keeps the probe 5cm from the oven area.

clean and free of foreign material. Readjust the door for proper closing.

- Do not defeat the interlock switches. If the oven operates with the door open, shut it down and check all interlock switches.
- Check the action of all interlock switches before and after repairing the oven.

- Before returning the oven to the customer, make sure the door is adjusted properly (tight and snug without any play).
- Never operate the oven when the safety interlocks are found to be defective.
- To meet the Department of Health and Human Services (HHS) guidelines, check the oven for microwave leakage with a recommended microwave radiation meter.

REQUIRED TEST EQUIPMENT

Besides an electromagnetic energy leakage monitor or other test equipment, a glass beaker and thermometer are required test equipment. Select a 600 or 1,000cc glass beaker or equivalent. Check with the manufacturer's literature for their required leakage tests. Most manufacturers recommend the glass beaker should hold from 8 to 10 ounces of tap water and be made of an electrically nonconductive material such as glass or plastic. A 100 °C or 212 °F glass thermometer is ideal for these tests. The water test is used when taking radiation leakage and normal temperature rise tests.

Place the water container in the center of the oven cavity. The placing of this standard oven load not only protects the oven, but ensures that any leakage is measured with accuracy. Set the microwave oven to high or full power. Close the door and set the timer for around three minutes. If the water boils before all leakage tests are completed, replace it with cool tap water. Now, turn the oven on for accurate leakage tests.

RADIATION LEAKAGE MONITOR

There are several survey instruments that comply with the required leakage procedure prescribed by the performance standard for microwave ovens. Those recommended by most oven manufacturers are: Holaday H1 1501, H1 1500, and H1 1800; Narda 8100, Narda 8200; Simpson 380M. Any one of the suggested leakage testers are accurate and will do the job in leakage tests. You may find some manufacturers request a certain leakage tester in their service literature. One thing to remember, the microwave leakage tester is no good unless you really know how to use it. Double check the meter leakage operation several times. Take several leakage tests to get the hang of it. You will find the leakage test is the last regular job in finishing the oven repair.

The Narda 8100 and the Simpson Model 380M are described thoroughly in chapter 2, Basic Test Equipment. Here, the operation and description of the Holaday H1 1500 is given. Although each model has its own method of operation, radiation leakage tests are the same. All of these leakage monitors operate on the 2,450 MHz band.

HOLADAY MODEL HI-1500 SURVEY METER

The survey meter contains a 1mA meter with an operational amplifier and is powered by two 9 volt batteries. You will find a fast and slow position switch. In the fast operation, the time for the meter to reach 90% of final value is less than one second. It takes just under three seconds with the slow switch position. A temperature compensation network for the diodes (within the probe) is located inside the meter box.

There are eight carrier diodes housed in the end of the teflon probe. This antenna array and diode detection has the unique feature of being able to sum the microwave electric field in a place perpendicular to the probe axis. The antenna is very broad and makes measurement easy when taking microwave leakage around the oven door and components. A 5cm cone spacer attaches to the tip of the probe. The probe is attached to the amplifier and meter box with a $3^{1}/2$ foot shielded cable.

Before you put the monitor into service, check the batteries and probe. Turn the selection knob to the battery test position. Make certain the meter hand reads above the green marker line on the meter. If not, replace the batteries.

To test the probe, turn the selector knob to probe test with the zero adjustment knob in the center position. The meter should read in the ok probe-test area between the green lines. If not, the probe may have been damaged and should not be used. Return the defective probe to the manufacturer. Do not try to fix it. All microwave leakage meters should be sent once a year to the manufacturer for correct calibration.

To operate the Holaday monitor, turn the selector knob to the desired meter range (2, 10, 100mW/cm^2). When the microwave oven is suspected of leakage, try the highest scale. Remember the maximum allowable leakage in the field is 5mW/cm^2. In most cases, the low 2mW/cm^2 scale will be used since leakage found in most ovens is very low.

Now, allow the meter to warm up under extreme temperature conditions. Place the tip of the cone against the equipment or component to be measured. The spacer cone keeps the detecting probe five centimeters from the equipment. Read the leakage level on the meter. Switch the meter to slow response when ovens with stirrer equipment are being tested. This makes the measurement easier to read. Always turn the selector knob off when the survey meter is not in use.

NORELCO RF RADIATION LEAKAGE TESTS

Check for leakage before and after servicing. If leakage is more than 5mW/cm^2, inform North American Phillips Corporation, attention product assurance department, immediately. After repairing or replacing any radiation material, keep a written record for future reference, as required by HHS regulation. This requirement must be strictly observed.

An electromagnetic radiation monitor, Holaday, Narda, or HHS accepted equivalent, 1,000cc glass beaker or equivalent, and a glass thermometer 100°C or 212°F are required test equipment.

Start by pouring 275 ± 15cc (9 oz $\pm \frac{1}{2}$ oz) of 20 ± 5°C (68° to 90°F) water in a beaker. Set the radiation monitor to 2,450 MHz and use it following the manufacturer's recommended test procedure to ensure correct results. When measuring leakage, always use a 2-inch spacer supplied with probe. Place unit in a cooking mode and measure leakage (using an electromagnetic radiation monitor) by holding the probe perpendicular to the surface being measured. Move slowly, no more than one inch per second.

Whenever the magnetron is replaced, measure for radiation leakage before the outer panel is replaced and after all the necessary components are replaced or adjusted. Special care should be taken in measuring around the magnetron and waveguide. These tests should be made with the outer panel removed.

With the oven fully assembled, including outer panels, measure all components, for radiation leakage around the door periphery, door viewing window, exhaust openings, and air inlet openings. Radiation leakage should not exceed the values prescribed below:

1. Maximum allowable rf leakage level is $5mW/cm^2$.

2. If the reading is over $5mW/cm^2$, refer to door and switch assembly adjustments section. The door is designed for leakage of less than $1mW/cm^2$ and adjustment should be attempted to bring all readings to below $1mW/cm^2$.

MEASURING PRECAUTIONS

1. Do not exceed full scale deflection.

2. Test probe must be moved no faster than one inch/second along test areas, otherwise a false reading may result.

3. Test probe must be held with grip portion of handle. A false reading may result if operator's hand is between handle and probe.

4. When testing near a corner of door, keep probe perpendicular to surface making sure probe end at base of cone does not get closer than two inches from any metal. If it does, erroneous readings may result.

5. When high leakage is suspected, do not move the probe horizontally along oven surface, this may possibly cause probe damage.

After adjustment and repair of any radiation-interrupting or radiation-blocking device, make a repair report, record measured values, and keep the data. Apply this measurement to the service invoice. If the radiation leakage is more than $5mW/cm^2$, first determine that all parts are in good condition and functioning properly, and that genuine replacement parts as listed in the manufacturer's manual have been used before notifying the manufacturer of the oven. Have the radiation monitor checked for accuracy by the radiation manufacturer at least once a year. Check the monitor batteries at least every six months or once a year.

WHERE TO CHECK FOR LEAKAGE

Microwave radiation leakage tests should be made around the front door and gap area (FIG. 11-3). When the magnetron tube or waveguide assem-

11-3 Microwave radiation leakage tests should be made around the front and door gap areas. Hold the test probe at the handle area, perpendicular to the cabinet door. Move the probe slowly across the gap area. Check the door for too much play, which may cause rf energy to escape.

11-4 Check for leakage around the magnetron tube. Go slowly around the seam area where the magnetron bolts to the channel cavity. Run the cone over the exhaust and air vent openings to check for possible leakage.

blies have been replaced, leakage tests should be made around these components. Always be careful while taking leakage tests around the magnetron with the oven in operation. Keep your hands on the handle area. Just let the plastic cone touch the various components.

Check around the magnetron tube for microwave energy leakage, before the outer panels are installed (FIG. 11-4). Go slowly around the seam area where the magnetron bolts into the channel cavity. Check around the heater or filament terminals for possible leakage. Double check around the magnetron if you find dented or bent areas in any of the metal areas.

Take a leakage test around the waveguide area seams (FIG. 11-5). Check for leakage on top of the waveguide cover. In a convection oven, if the shielded metal areas have to be removed to repair the convection heating elements or to get access to them, check for leakage after these components have been replaced. Try to avoid the belt driven pulley and fan rotating components while taking these tests. You should not encounter hv terminals on the top area of the oven, but do be careful of possible low-voltage connections.

11-5 Take a leakage test around the waveguide seam areas. Go completely around all possible seam areas and check for leakage. Slowly check these areas when the waveguide cavity has been replaced.

Always take a leakage reading around the front door area after the oven has been repaired and cleaned up. When testing near a corner of the door, keep the probe perpendicular to the surface making sure the probe end at the base of the cone does not get closer than two inches from any metal. Slightly pull on the door while the oven is operating and notice if there is any leakage. Actually, the door should have no play, letting the microwave energy escape. This test may show up a possible damaged rubber or choke area. Sometimes, if the door is warped you may try pulling out on the door handle to uncover a leakage spot at the corner. Usually, a defective door will leak at the top and bottom corner areas.

Another method to check the door is to slightly push down on the door latch and take a leakage test. Do not push down too hard or the latch will release the door, stopping the oven. Just press down slightly before latch interruption. Now, take a leakage test around the end and corner door areas while holding down the door latch handle (FIG. 11-6).

11-6 Check the door for leakage by pulling on the door or pressing down on the door latch. Now, test for leakage around the end and the corner door areas while holding down the door latch handle. Do not press too hard, or the door may open and stop oven operation (courtesy of Matsushita Electric Corporation of America).

Check for leakage around the door viewing window after testing around the door area. Double check all exhaust and air inlet openings. You may find them on the top or both sides of the microwave oven. Make sure the screws inside the oven front door are not loose or missing. Microwave energy may leak if the screws are not properly tightened. Slowly approach these areas for possible high leakage. When high leakage is suspected, do not move the probe horizontally along the oven surface or you may cause probe damage.

TIPS ON TAKING LEAKAGE TESTS

Do not exceed full-scale meter deflection. Slowly approach the tested area for signs of high-leakage radiation.

Go slowly—take your time in taking leakage tests around the door. Proceed at no faster than one inch per second (2.5cm/sec) along any suspected area.

Make sure the plastic cone is slipped into place and keep the probe perpendicular to the testing surface.

Keep your hands on the probe handle to prevent a false reading. Slightly pull out on the oven handle and then take a leakage test.

Always handle the leakage tester with extreme care. Keep the delicate test instrument in its case when not in operation.

Most manufacturers recommend the leakage tester be sent back in for calibration at least once a year. If in doubt about a leakage reading, go back over and check it once again.

Be careful when taking leakage measurements after replacing the magnetron, when there is cabinet damage, or when replacing the front door. You may encounter excessive rf radiation. If the instrument indication goes to full scale at any time, the tester may be damaged due to the presence of excessive microwave power.

Always approach the oven slowly with the probe while observing the meter. The oven should be set at maximum power output. Hold the probe two or three feet from the oven door. Move slowly toward the door, oven surface, or gap between the door while watching the meter. When high leakage is noticeable, do not move the probe along the oven surface, you may burn out the probe. Often the greatest leakage is at the corners of the door. Proceed with the leakage test when excessive leakage is not found on the oven.

TYPICAL LEAKAGE TESTS

The test probe must be held by the handle to avoid the false reading of the test instrument. Hold the probe perpendicular to the cabinet door or component to be checked for radiation. To check the door, run the probe cone on the door and cabinet seam. Move the probe slowly (not faster than one inch per second) along the gap area while watching the meter. Check for the highest or maximum indication on the meter.

Keep the probe perpendicular to door corners or access area so the probe will not be closer than the cone space (two inches). Always use the two-inch spacer with the probe. Often leakage occurs around the door corners. Double check the door area. When any leakage is noted in the top center or end areas of the door, push tightly on the door to determine if the door has some play between it and the oven. If the leakage reading decreases by pushing on the door, check for door adjustment. The door must be tight against the oven seal.

After the door is adjusted, take another reading. Usually the long door latches do not seat tightly in the oven or microswitch area. Adjust the door latch head at a position where it smoothly catches the latch receptacle through the latch hole. You may have to reset or adjust the monitor and interlock switch assembly. Adjust the door to the manufacturer's specifications. Check for leakage at the door screen and sheet metal seams.

Take leakage tests around the magnetron tube after installing a new tube. Check for possible leakage between the magnetron and the waveguide assembly. Run the cone over the exhaust and inlet air openings for possible leakage (FIG. 11-7). The door and all cabinet openings should be checked for leakage after the back cover has been replaced. You may find most ovens have no or very little leakage (under 0.1mW/cm²).

11-7 Don't forget to check for loose screws or panels at the front door for leakage. Make sure all screws are tight. Besides causing leakage, a loose screw can cause arcing in the oven.

Record all leakage readings of any oven on the repair ticket for future references. Although the customer does not understand the leakage reading, they want to know if the radiation will cause damage. Most manufacturers want the submitted warranty report to include the oven leakage, while others require the make and model number of the leakage tester to be recorded. When the leakage is over 5mW/cm², report the leakage reading to the manufacturer at once.

CUSTOMER LEAKAGE TESTER

There are two readily available microwave leakage detectors on the market. The GC-20-224 model is a small round instrument at a cost of $14.95 (FIG. 11-8). Radio Shack has a microwave leakage detector Model 22-2001 for less than $15.00. Although these two detectors are not certified leak-

11-8 A low-cost leakage tester can spot potentially dangerous rf leakage in a faulty microwave oven. It's only used by the customer for possible safety checks.

age instruments, they may spot potentially dangerous rf leakage from a faulty microwave oven. They may satisfy the housewife who is afraid of possible microwave radiation.

The GC micrometer radiation leakage detector may be operated like any leakage tester. Measure one cup of cold water and place in the microwave oven. Close the door and set the cook selector timer on high for 60 seconds or longer. Also, the leakage test may be made while cooking food. Hold the meter perpendicular to the oven door edges, like the cone probe of any leakage tester. Move the meter at approximately two inches per second. If the meter stays in the green zone, the emission level is considered safe. Any time the needle enters into the red zone, the oven should be checked with an authorized microwave leakage tester or at a microwave service center.

The Micronta microwave leakage detector from Radio Shack is easy to use (FIG. 11-9). Simply run the end of the test instrument along the seams of the door or intake and outlet openings and watch the meter movement. There are only two areas to watch on the meter. If the meter hand stays in the green area, no relative amount of microwave energy is escaping. When the meter hand goes into the red scale, call your microwave oven repair technician. Just remember these two low-cost microwave leakage instruments are not certified, but may prevent possible serious and even permanent health problems.

LEAKAGE RECORD KEEPING

After adjustment and repair of any microwave energy or microwave energy blocking device, record the leakage reading on the service

11-9 A Micronta microwave leakage detector from Radio Shack is another type of radiation meter. Simply run the test instrument along the seams of the door or the intake and outlet vent areas. If radiation leakage is detected with one of these instruments, have the oven checked with a certified radiation energy survey meter.

invoice. Also, don't forget to place the leakage reading in the service literature. When making out a warranty repair report, most manufacturers ask for a leakage test recorded on the warranty report. This report may take the repair agency and manufacturers off the hook, if for some unknown reason a customer tries a lawsuit for radiation burns.

Always report to the manufacturer if the microwave leakage may be more than $2mW/cm^2$ ($1mW/cm^2$ for some ovens in Canada). Don't let the customer have the oven until the manufacturer releases the oven. Follow the manufacturer's instructions in repairing or replacing the microwave oven. The factory may want the oven shipped directly to them for possible repairs.

REPLACING THE DOOR

Check the door for warped or damaged areas when excessive radiation leakage is noted at the top or corners of the door. In most cases, the door is opened with the handle or an extra hand at the top of the door. Eventually the door becomes warped or may be damaged while it is open. A broken or worn hinge may let the door sag creating possible leakage. Inside oven explosion of food may cause the door to be blown open causing damage to the door. In some ovens, you may find a broken latch-head assembly not holding the door snug against the open face plate. Definitely replace the door assembly when the front glass is cracked and broken or the door cannot be properly adjusted.

Inspect the choke and door seals at the inside of the oven door. Check for damaged areas. Make sure that all inside door screws are not loose or missing. In some doors the glass may be removed and replaced. Usually, the whole door assembly is replaced when damaged or warped out of line.

After determining the door assembly should be replaced, check the parts list for the correct part number. Always list the correct part and model number when ordering any oven component. All oven door components must be ordered directly from the oven manufacturer or the manufacturer's distributor.

A typical oven door may be hinged at the top and bottom or have a long piano-type hinge down the entire backside of the door (FIG. 11-10). A hinge plate over the piano-type hinge assembly must be removed before the door is free. The new door should be replaced in the reverse order of removal. When only two pivot area hinges are found, the top hinge holds the door towards the oven while the bottom hinge supports the door load.

11-10 A typical oven door may be hinged at both the top and the bottom or have a piano-type hinge down the entire door side. Here is an oven with an offset hinge. This door can be adjusted by loosening the screws in the hinge and moving the metal plate behind the hinged area.

Make sure the new door is tight and level with the front piece of the oven. Check the latch lever or door hook alignment pin for smooth operation. Some doors are held tight to the front piece by correct adjustment of the latch interlock assemblies. Others are adjusted at the door hinge areas. It's best to follow the manufacturer's literature for correct alignment.

Although most oven door hinges are held in place with hinge mounting nuts or Phillips screws, you may encounter in some new ovens a new Torx screw. These screws resemble the Phillips head except the Torx screw has a star type indentation. Do not try to remove or tighten these Torx screws with an ordinary screwdriver. Pick up a Torx screwdriver, size T-20 or T-28, at your local hardware store.

Remove the outer cover of all ovens before attempting to adjust or remove the door. Then discharge the hv capacitor. Here are several typical microwave oven door replacements and adjustments.

Amana Model RR-40

Open the door wide open. Remove two counter-balance mounting screws from each side. Lift off the door. In this model the door handle may be a separate replaceable item as well as the door glass. Adjust the level and tighten the two mounting screws after door replacement.

Hardwick Model EN-228-0

Remove the cabinet trim from the main front by removing several screws on top, bottom, and sides. Lay the oven on the backside and remove four hex head nuts that secure the door hinge to the main cabinet. Lift door handle and door from the base unit.

To install a new door, reverse the above procedure. Level the door and keep 1/16-inch spacing between the control panel and door edge. Tighten the hex nuts starting at the top and ending at the bottom area. Open the door and notice proper door alignment. Loosen hex nuts and readjust the door, if needed.

Norelco Model MCS 7100

Turn the oven on its backside. Loosen the upper hinge mounting nuts and remove the door assembly. Be careful not to lose the small nylon spacer at the top and bottom hinge. Make sure both nylon spacers are in place when installing a new door. Loosen lower hinge mounting nuts (FIG. 11-11).

Now, press the door against the oven face plate near the hinge and retighten the hinge mounting nuts. Set the oven dial for normal operation. Check the play in the door. Readjust the door interlock switches if there is too much play. Loosen the two side mounting screws of the metal mounting bracket of the latch switch assembly. Press in on the door and pull back on the latch assembly. Tighten the latch bracket mounting screws. Check the door for proper clearance. Make sure the upper and lower door switches are functioning properly.

11-11 You can adjust the door removal on a Norelco Model MCS 7100 oven by loosening the two hex nuts at the top and bottom door hinges. Make sure both nylon spacers are in place when installing a new door. Level and align the door, and then tighten all four hex nuts (courtesy of Norelco Corp.).

Quasar Model MQ55-207W

Remove the two hex nuts holding the upper hinge to the oven base. Open the door and pull the door towards you. Be careful. The top door arm must come through the slot in the oven assembly. Leave the bottom hinge intact. Lift the door up and off the bottom hinge pin.

Before installing the new door, check for a washer spacer on the bottom hinge. Adjust the door parallel to the cabinet. The door should be tight against the cabinet base with no clearance. The top and bottom hinge may be adjusted to align the door correctly with the control panel. Now, tighten all four hex nuts to secure the hinge and door assembly. Open and close the door for proper operation.

Samsung Model RE-705 TC

Remove the small spring attached to the door arm cam. Remove the rod pin that prevents door from flying open. You will find two metal hex screws at the top and bottom hinge. Remove them and pull the door out. Be careful when pulling the door arm cam through the slotted area.

Install the new door with the reverse procedure. Adjust and align the door parallel with the control board assembly. Keep the hex nuts loose at the top and bottom hinge until the door is in place. These hinge brackets can be pushed back and forth and to the side for proper adjustment. Now, tighten up all four hex nuts (FIG. 11-12). Make sure the door is level and closes freely.

11-12 Tighten all four hex nuts for proper door adjustment on a Samsung Model RE-705TX. Make sure the door is level and closes freely.

Sharp Model R-7810

Lay the oven on the backside and remove four metal screws of the piano-type hinge assembly. The metal back-strap plate will come loose. Lift the door up and off the base assembly. Be careful not to damage the control panel assembly.

Install the new door in the reverse procedure. Make sure the door is parallel with the bottom line of the oven face plate. The latch heads should pass through the latch holes without binding. Keep the door tight against the oven face plate. Insert all four metal screws. The door may be aligned before the screws are tightened in the long hinge assembly. Now, check the door for proper clearance and operation.

After each door replacement or adjustment, double check around the door for poor alignment and leakage. Try the door several times in opening and closing operation. Now, go around the door and check for possible radiation leakage with a radiation leakage meter. If any leakage is noticed, realign and adjust the door.

The door on a microwave oven is designed to act as an electronic seal to prevent microwave energy leakage from the oven cavity. If light can be seen on some point around the door, it's still normal, provided the radiation meter does not show any leakage. A normal door may not be air tight or moisture tight. When light movement of air or moisture appears around the oven door, this does not mean the oven is leaking radiation energy. When these conditions occur, double check for leakage with the radiation monitor survey meter.

Some electronic technicians use a dollar bill between the door and oven to test the snugness of door. If the door is proper and tight, the bill cannot be pulled easily out of the door. Realign the door if the dollar bill is easily pulled out.

REPLACING WAVEGUIDE ASSEMBLIES

If the waveguide has been removed or replaced with a damaged waveguide assembly, or when replacing the stirrer motor or antenna, make sure the waveguide bolts and nuts are tight (FIG. 11-13). Visually inspect any high places in the waveguide assembly. Remove and check for particles under the metal waveguide. If the waveguide is not tight, rf hot spots may occur with possible rf leakage. Double check the waveguide assembly with the radiation monitor probe.

LAST MINUTE CHECKS

- Do not allow oven to be operated with the door open, or improper interlock operation.
- Check all interlocks, proper door closing, wear and tear of rubber or choke seals, damaged door hinges or latches, and damaged door windows.
- After completing all repairs, check the magnetron, waveguide, and

11-13   Inspect the waveguide assembly for hot spots and rf energy radiation. Run the radiation monitor along the waveguide metal seams.

oven cavity for proper alignment, snug bolts and nuts, and good connections.

- Repair and replace all components with exact oven part numbers.
- Check for proper fuse. Do not clip across fuse terminals to make oven operate. Replace defective fuses with those of proper amperage and type.
- Make microwave leakage tests with approved radiation monitor meters required by federal laws after all repairs are made. Record the measurement on service invoice.

Chapter **12**
Microwave oven case histories

*I*n this chapter you will find actual microwave oven troubles that have occurred during my years of servicing microwave ovens. Although each service problem may be somewhat different, you may find they are very similar to the type of problem you are now experiencing. Since many ovens are manufactured for other firms, the different units may be identical in operation. These actual case histories were chosen to help locate the various service problems found throughout microwave ovens. Check the various microwave oven symptoms and compare the problem to the oven being serviced.

NOTHING WORKS
K-Mart Model SKR-6705—Dead oven

The fuse was replaced in this oven and the light came on. When opening and closing the door, the fuse blew again (FIG. 12-1). Replace the monitor switch, QSW-M0046Y BEO. This oven is manufactured by the Sharp Corporation.

K-Mart SKR-7805—Dead—No light

Replaced 15 amp fuse. Checked monitor and interlock switches for correct operation. Readjusted oven door.

Litton Model 70/05 830—Dead

Checked the door for too much play after replacing the 12 amp fuse. Adjusted the latch switch and door for proper closing. Sometimes if there

12-1 The 15 amp fuse provides overloading protection of the oven circuit. Check the interlocks, magnetron, triac, varistors, and hv diodes for most fuse-blowing components. The fuse should show a dead short on the DMM.

is too much play between door and base area, the monitor switch may hang up causing the fuse to blow once again.

Litton Model 70-05—Dead

Replaced fuse and installed a new latch switch. While inside the oven, checked for dark or dead light bulbs. This model happens to be a commercial microwave oven. Checked for defective on/off switch. This switch may be located on front of oven or at top or side (FIG. 12-2).

Norelco Model RK-7000—Dead

After checking the fuse and voltage to and from the control board, a new board was ordered. Replacing the entire control board solved the dead condition. Replaced only the control board, not the push button assembly (part number 6000-000-00200).

Norelco Model MCS 7100—Dead

When the oven door was closed, the oven appeared dead. Again check those interlock switches. Replaced interlock switch 6100-000-0055. Check for curled up plastic over switch terminal connections, indicating poor switch contacts.

12-2 Sometimes the on/off switches actuated by the front door may have poor contacts or are bent together. They can be mounted behind the front panel or alongside the oven. Here, the ac switch is located on the top left hand corner of this
Samsung Model RE-705TC.

Norelco Model MCS 8100—Dead—Blown fuse

After replacing the fuse, the oven began to operate, except for fan rotation. Found the fan assembly screws had come loose, letting the fan blade
jam. The fan motor would overheat and blow the 15 amp fuse. Replace
the bolt and nuts. You may find them in the bottom oven area.

Norelco Model MCS 9100-Dead

Control board assembly will not count down—not even a fan operation.
Replace control board assembly 8100-000-00008.

Norelco Model MCS 8100—Dead—Nothing

Removed outside cover and replaced blown 15 amp fuse. Checked the
fuse continuity with the low-ohmmeter range of the vom. Found a loose
plug going to the power control board. Repaired plug and terminals
before starting up the oven.

Quasar Model MQ6620TW—Dead—No control function

Checked control board. Would not count down or set correct time.
Replaced entire control board (part number 84-90 564A41).

Samsung Model RE-705TC—Dead—Check leakage

Replaced 15 amp fuse. Checked interlock switches and adjusted up door. Ran four hours cooking—water test. Check the door for radiation leakage. Customer had complained of loose door and possible leakage.

Sharp Model R6770—Dead

In some of the Sharp Models that have a rotating temperature control, the dead condition may be improper control setting. Check the temperature control for off condition. If the temperature probe is unplugged, the temperature control should be set in off position. Many times the ovens come in for repair when the operator forgets to turn the temperature control off from the temperature probe cooking operations.

Sharp Model R7650—Blown fuse

After replacing the 15 amp fuse in this model, check the monitor switch and primary interlock switch. Replace the monitor switch if keeps blowing fuse when opening the door.

Sharp Model R7650—Dead—Lights on

The fuse was good since the oven lights were on but there was no cooking operation. Check each interlock switch after discharging the capacitor. Notice if the switches may have black or burned marks at the terminal connections. Clip leads across each switch terminal and monitor with a voltmeter. If the switch is open, you will have the power-line voltage (120 Vac) across these terminals. Make sure you have the right switch or interlock. In this case, the secondary (upper) interlock switch was open (QSW-M0055YBEO).

Sharp Model R7650—Oven lights—No operation

The oven lights were on which meant the fuse and timer were operating. Checked interlock and cook switch, they were normal. While checking the door, found a broken latch spring. The plastic latch hook was down and would not trigger or press on the interlock switch. These latch springs must be replaced in the oven door (FIG. 12-3).

Sharp Model R7650—Dead—Won't turn on

Replaced the 15 amp fuse. Checked upper interlock and monitor. Replaced both for erratic operation (FIG. 12-4).

Sharp Model R7650—Oven light on—Dead

Nothing happened when the cook button was pushed. Replaced the interlock switch.

12-3 In some ovens, the oven light is turned on with a switch mounted at the front corner of the microwave oven.

12-4 Any latch switch can be checked with the low-ohm scale of a vom. Clip the leads to the switch terminals. Always pull the power plug and discharge the hv capacitor before taking resistance measurements. Now, open and close the door. A low reading will result when the contacts are closed, and there will be no reading when they are open.

Sharp Model 7600—Dead—Door adjust

The customer complaint was that you have to hold open the door to make the oven come on. When this condition occurs, realignment of the door and switch assembly adjustment should be made. Always check the door for possible radiation leakage after door adjustment.

Sharp Model R7650—Dead

Replace the open 15 amp fuse. Check all interlock switches for proper operation. Replaced defective interlock switch QSW-M0036YBEO.

Sharp Model R7650—Dead—Won't come on

Replaced spring in door latch and also replaced top interlock switch (part number MSPRD0013YBEO).

Sharp Model R7710—Dead

After replacing blown 15 amp fuse, checked all interlock switches. Found monitor switch was sometimes hanging up. Replaced both monitor and secondary interlock switch.

Sharp Model R9200—Dead

In this model both latch springs were broken inside the oven door. The latch hooks will tie down and not engage the microswitch of the interlock switches. Replace both with part number LSTPP00179BFO.

Sharp Model R9500—Oven won't come on

The fuse checked normal. Voltage was going into the control board but not control function. Replaced control board number BUNTK085-DE00.

Sharp Model R9600—Dead

Replaced blown fuse. Found door had too much play. The oven door should be snug against front piece of oven. Pulled back door latch assembly and tightened up assembly. Sometimes a drop of cement on screws will not let them loosen up in operation.

Sharp Model R9750—Dead

Fuse okay. Found a loose plug in connection on the readout control board. Repaired and rechecked oven. Bench check time took four hours of intermittent water test.

Toshiba Model ER749BT1—Dead

Replaced blown fuse. Replaced 5 amp monitor switch.

OVEN LIGHTS OUT

Sharp Model R7704—Lights out—Cooks normally

Removed outer back cover. Discharged high-voltage (hv) operation. Found lamp bracket had come loose. Replaced and installed a new light bulb RLMP0004YBEO.

Sharp Model R7704A—Light out and noisy

In this model, replaced both light bulbs. One was completely out and the other very dark, indicating only a few more hours of operation. Found loose bracket on cooling fan. Replaced metal screw and ran cooking test.

Sharp Model R8200—No oven light—No operation

Straightened top oven arm. The arm was bent out of line and would not trigger primary interlock switch. Replaced dead oven bulb.

Toshiba Model ER749BT—Flashing oven light

This oven operated correctly with an intermittent oven light. Removed both covers. Found light bulb loose in the socket. Tightened bulb and checked all interlock switches.

TIMER ROTATED BUT NO OPERATION

Litton Model 37000—Timer on—No operation

The defrost or timer motor would not run in this microwave oven. Found the bottom door interlock switch not making contact. The switch assembly had slipped back and was not making contact. A new interlock switch was installed. Secured the switch assembly with a dab of glue on each nut to keep the assembly in place.

Sharp Model R6750—Intermittent—Now dead

In this model the timer was not dependable. Sometimes the timer was too slow and other cooking modes too fast. Finally, the oven went dead. The 15 amp fuse was replaced. The sluggish timer assembly was replaced (FIG. 12-5). After replacing the new timer, a four hour cooking and leakage test was made before releasing the oven.

12-5 A sluggish timer can be used with a worn or binding gear assembly. To check a dead timer, measure the voltage across the timer connections. Check the switch connection for open resistance measurements.

Sharp Model R7704—Intermittent timer

This oven was timed with a watch and appeared intermittent. A whole new timer unit was installed. In some of these bell-type timers, the end of the cooking cycle may sound like a thud instead of a clear bell sound. Adjustment of the bell clapper may be made by bending arm in or out.

Sharp Model R7600—Light—No timer action

When the timer was rotated, the blower motor or cook light did not come on. Found a defective interlock switch. All interlocks were checked for proper operation before cooking tests.

Sharp Model R7600—Intermittent

Front door hook would not trigger the microswitch. Bent latch hook in place and ended up replacing interlock switch assembly. If door is too loose, pull interlock switch assembly back and tighten securely. Make sure all latch hook assemblies will trigger interlocks.

Sharp Model R8200—Timer off

The timer ran very slow in this model and sometimes would almost stop. Several cooking tests were made before a new timer assembly was installed. This timer assembly must be ordered directly from the manufacturer or manufacturer distributor (FIG. 12-6).

LIGHT ON BUT NO FAN OPERATION

Litton Model 370.00—No fan or defrost motor operation

This oven light would come on but no fan operation and no cooking mode. Suspected bad timer. Found a defective latch or interlock switch

12-6 This Sharp timer assembly was installed to replace one that was very slow. Check the continuity of the timer motor and the switch contacts for a dead timer operation.

had slipped out of position and latch arm would not engage switch. Replaced and remounted interlock switch. Cemented switch into position.

Norelco Model MCS 6100—No fan or timer action

Replaced upper primary interlock switch assembly. This switch is mounted on a metal strike switch assembly and is held in place with metal screws. Check all interlock switches. Open and close the door several times while in the cooking mode to make sure all interlock switches are working.

Norelco Model MCS 7100—Bad latch switch

With no fan, no cooking action, the upper strike switch assembly was replaced (part number 6100-000-00055). The upper primary and lower secondary door switches are mounted on a latch slide-assembly which moves up when the door is closed. As the slide moves up, the switches are activated by latch hooks mounted in the door. Be sure and line up the adjustable strike assemblies with the door hook, by simply loosening the screws on the strike switch assembly and tightening the screws into position (FIG. 12-7).

12-7 Some ovens have a circuit diagram glued to the inside tip panel cover of the oven for easy troubleshooting.

Norelco Model MCS 6100—No fan operation

Check for a broken lead connection of the fan motor assembly (FIG. 12-8). With some ovens the fan motor may be plugged in and the plug will vibrate, causing the plug to work loose, disconnecting the ac to the fan motor. Measure the ac voltage across the field terminal of the fan motor (120 Vac). If voltage is normal with no fan motion, check the motor winding with the ohmmeter. Replace fan motor assembly.

12-8 After inspecting the fan motor lead for possible breaks or bad connections, measure the voltage across the fan motor field wires (120 Vac). The continuity of the motor may be checked with the low-ohm scale of a vom.

FLICKERING LIGHTS

Amana Model RS470P—House lights flicker

The oven was suspected of drawing heavy current until the microwave oven was found operating from an extension cord. The technician requested that an electrician put in another grounded circuit from the fuse box to the microwave outlet.

Litton Model 402.000—Flickering cook light

The customer complained when the oven was operating the cook light would constantly flicker. Most ovens with neon bulb type cook lamps will flicker to some extent. These neon bulbs have a voltage dropping resistor and operate from the 120 volt power line (FIG. 12-9). When the bulb becomes very dim, check the bulb for dark areas of the glass envelope. Replace the neon bulb. Sometimes cleaning off the grease and dirt will help the brightness of these small neon bulbs.

12-9 Check the neon cook light for a burned area or flickering light. Usually, a voltage-dropping resistor is in series with the neon cook light. The cook light is mounted right in the switch area.

Sharp Model R7650—Light flickering, door is open

Check for a dirty or defective secondary interlock switch when the oven light continues to flicker with the door open. First, tighten all light bulbs in their respective sockets. Replace upper secondary switch with part number QSWMC055Y-BEO.

Norelco Model MCS 8100—Lights on—No fan

No fan or cooking operation was noted in this oven. When the door was moved, the oven began to operate. The door was adjusted and both interlock switches were adjusted for correct microwave cooking operation.

Sharp Model R6770—Lights only—No operation

The only problem with this oven, the operator had turned the temperature probe on and was not cooking with the probe. When set in the on position, the oven will not cook with probe out. Showed the operator how to run oven in the off temperature probe position with normal cooking procedures.

Sharp Model R7650—Lights—Dead operation

Check all interlock door switches. The secondary upper interlock switch was defective. Adjust the interlock switches so the door fits tight against the oven base area.

Sharp Model R7704A—No fan operation

Here a badly soldered connection was found on one of the motor terminals. If arcing has occurred, remove the wire, scrape with a pocketknife and tin with rosin solder paste. Reconnect the terminal wire and solder it.

Sharp Model SKR7705—Fan quits

The fan motor would stop rotating after the oven was operating for several minutes. With the oven unplugged and the hv capacitor discharged, the fan blade was rotated by hand. The blade would barely move. Washing out the fan motor bearings and proper lubrication solved the intermittent fan rotation (FIG. 12-10).

COOK LIGHT ON BUT NO TURNTABLE ROTATION

K-Mart Model SKR90505A—Turntable won't turn

The turntable within the oven cavity would not turn in this model. Turn the oven on one side to get at the turntable motor assembly. The turntable motor was removed. Glancing at the motor field coil indicated the motor had become very warm.

12-10   Most fans are located where you can easily see them. Here is an unusual squirrel-cage fan blower in a Samsung PE-750TC model.

The motor field coil was charred and a new turntable assembly was ordered (ROMTE0061Y BEO). A broken plastic coupling from motor to turntable had bound the motor so it could not rotate. The turntable motion was restored with a new motor and coupling assembly FCPL0018WRKO.

K-Mart Model SKR7805A—No turntable rotation

All other oven functions were perfect except for no turntable rotation. The motor turntable was removed and found a broken plastic coupling. Replacement of the plastic bushing (FCL-0018WRKO) cured the no turntable rotation.

Sharp Model R7600—Turntable quits

The turntable would quit rotating after two or three minutes of cooking. All cooking modes were normal except the turntable rotation. The turntable coupling was okay and the motor assembly was removed. The jammed gear box was replaced with a new turntable assembly (part number RMOTEDO37Y BEO).

Sharp Model R7600—Fan motor on—No oven light

Here the turntable motor was used as a trouble indicator. The turntable did not rotate nor was there a cooking mode. The fan motor was rotating but no oven light. The defective component turned out to be a defective interlock on the left-hand side of the cabinet.

This switch assembly can be pushed back and forth for adjustment after loosening two metal screws. The interlock switch must be triggered with a long arm from the bottom of the door with the door closed. When the door is open, the ac switch contacts are open. After correct adjustment, tighten the two screws securely. Drop cement on the metal screws to hold them in place.

Sharp Model R7700—Intermittent turntable rotation

With this model, the turntable may operate for several cooking modes and then stop rotating in the middle of a long cooking session. A broken plastic coupling was replaced after dropping the turntable motor down for observation. Check the turntable motor and gear box for correct lubrication.

Sharp Model R7704A—No turntable rotation

Found the turntable would not operate with all other oven functions okay. Found a broken and jammed plastic bushing. Replaced plastic bushing with part number PCPL 0018 WRKO. This same coupling part number is used in all turntable assemblies.

Sharp Model R9700—No turntable rotation—Dead oven

The turntable motor was used as a trouble indicator. Besides no turntable rotation, the whole oven was dead. The 15 amp fuse and interlock switches were normal. Replacing the defective control unit BUNTK085LDE00 solved the no turntable rotation problem.

NORMAL OVEN OPERATION BUT NO HEAT OR COOKING

K-Mart Model SKR-7705—Oven light—No cooking

The fan with all other functions seemed normal except no heat. High voltage and current readings were monitored at the magnetron. Very low high voltage and no current were observed. Removed heater leads from the magnetron with very low high voltage.

Discharged the hv capacitor. Made resistance test between hv diode and chassis ground—resistance below 2 kilohms (FIG. 12-11). Removed defective hv diode and replaced. Made another voltage and current check before cooking water test.

K-Mart Model SKR9505—Lamp out—No cooking

Normal fan motion. Replaced burned out oven lamp. Found high voltage at the magnetron but no current. In fact, the high voltage was higher than normal, indicating an open magnetron. Installing an FV-MZOY 11WRKO magnetron solved the no heat problem.

Litton Model 402.000—Blown fuse—No heat

After replacing the 15 amp fuse the oven was normal except no heat or cooking. No high voltage was found on the heater terminal of the hv mag-

12-11 In the K-Mart SKR-7705 oven, a resistance of 2k ohms was found across the hv diode. If a faulty diode is suspected, remove one end of the diode and take another resistance measurement.

netron. Always discharge the hv capacitor to measure continuity between magnetron and chassis ground.

Only 1.5 ohms was found between heater terminals and ground. The short was still present when the heater terminals were removed. The magnetron must be shorted internally. In this model, check the hv diode, which is mounted down inside the magnetron shield area (FIG. 12-12). Removing one end of the diode resulted in a shorted hv diode. One could have easily ordered a new magnetron with only the replacement of a shorted hv diode being needed.

Litton Model MD 10D51—No heat—Defective capacitor

Everything operated except no heat or cooking in the oven. The hv capacitor was discharged and no arc was noticed between the two capacitor terminals. No high voltage or current readings were noticed on the magnetron. Undoubtedly, the hv circuits were defective.

Resistance readings were made between the hv diode and ground

12-12 Most hv diodes are located out in the open. However, in this Litton Model 402, the diode is located inside the magnetron shield.

with no results. The resistance reading of the hv and heater windings were normal. A 10 megohm reading between heater and chassis ground was above normal. Power-line voltage (120 Vac) was found on the hv transformer.

Since no high voltage was measured at the magnetron, the hv capacitor was suspected. Two alligator lead clips were clipped across another hv capacitor and then placed across the suspected capacitor leads. The high voltage came up with correct magnetron current. Replacing the open hv capacitor provided adequate heat in the oven.

Magic Chef Model MW207-4—No cooking

No high voltage was measured at the anode side of the hv diode. Resistance measurement of both transformer windings were normal. Power-line voltage was measured on the primary winding of the hv transformer. The resistance measurement between anode of the hv diode and ground was about 10 megohms. Either the hv diode or capacitor was defective.

A new hv capacitor was clipped across the capacitor terminals with no high voltage. The cathode end of the hv diode was disconnected from chassis ground. Just clipping a new diode into the circuit solved the hv problem. Be careful while working around the hv components. Discharge that capacitor each time the ac plug is pulled (FIG. 12-13).

Magic Chef Model MW207-4—No heat—No high voltage

In another Magic Chef microwave oven the results were the same—no high voltage. Of course, the hv problem could easily be seen. The cath-

12-13 The hv diode should be located near the hv capacitor terminals in newer microwave ovens. The hv diode should be located near the hv capacitor.

ode lead of the hv diode was broken right in the body of the diode. A new hv diode brought back the no heat—no hv problem. In some cases, when the diode lead is broken off and is long enough, another wire may be soldered to it instead of installing a new replacement.

Montgomery Ward Model KSA-8037A—No cooking

Low high voltage was noted upon the heater terminals of the magnetron with very little current drain. The hv capacitor was discharged and a resistance reading was taken between the heater terminals and chassis ground. A 5 kilohm measurement indicated a leaky magnetron or hv diode.

The cathode end of the diode mounted to the chassis was removed. A resistance reading across the diode could not be measured. No doubt the magnetron was leaky. Sure enough, a 5 kilohm reading was found from the heater terminal to the metal shield of the magnetron (FIG. 12-14). Installing a new magnetron FV-MZ0033YBKO solved the no cooking operation.

12-14 A 5k ohm reading from the heater terminals to the metal shield of the magnetron indicates a leaky diode or magnetron. This magnetron had internal leakage and was replaced.

Montgomery Ward Model L65-10/47X—No heat—No cooking

An hv measurement was above normal in this Wards microwave oven with no current reading. The hv capacitor was discharged. Both heater terminals were removed. The heater terminal transformer winding showed continuity. No resistance reading was noted between the heater terminals of the magnetron. Here less than 1 ohm should be measured. A new magnetron was installed.

Montgomery Ward Model 67X175—No cooking

No high voltage or current was found on the magnetron tube. Discharge the hv capacitor before taking resistance measurements. When taking resistance measurement across the hv diode, a bubble type bulge was found on one side of the diode. Sure enough, the hv diode measured less than 1 kilohm. Sometimes when the hv diode or capacitor becomes leaky, the 15 amp fuse may open. It depends on how long the oven is on with a leakage across the hv transformer.

Norelco Model R6000—Won't heat

Very low high voltage and no current measurement was found on the heater terminals of the magnetron. Undoubtedly, a leaky magnetron or hv components were involved. The hv capacitor was discharged with no arcing. A 3.5 kilohm reading was found across the heater terminals to ground.

Both heater terminals were removed. The resistance measurement from heater to ground was above 10 megohms. A continuity reading was measured between the two heater terminals. A 3.5 kilohm measurement was found across the hv diode. Even the body of the diode was warm. Sometimes after discharging the hv capacitor, you may find a defective diode by feeling how hot the body is. Some of these diodes are warm after several hours of operation and this is normal. An overheated diode should be replaced.

Norelco Model MCS 6100—No heat—No cooking

High voltage was normal with no current reading in the microwave oven. Continuity checks on the hv and heater windings were quite close. When high voltage is present with no current reading, you may assume the magnetron is open. Sure enough, the heater terminals were open on the magnetron (FIG. 12-15). A new magnetron was installed (6100 00 00029).

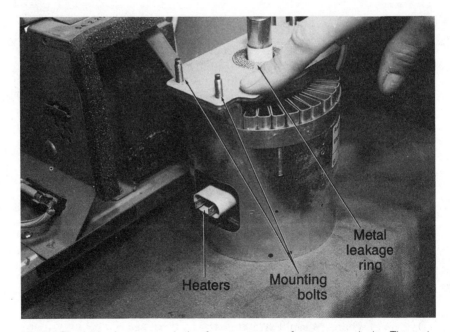

Heaters   **Mounting bolts**   **Metal leakage ring**

12-15 Check the heater terminals of any magnetron for open continuity. The resistance between heater or filament terminals should be less than 1 ohm on a low-ohm scale of vom.

Norelco Model R6008—Normal, except no heat

No hv or current reading was found on the magnetron. Regular power-line voltage was found on the primary winding of the hv transformer. Both hv and heater windings were normal. A resistance measurement of 1.5 kilohms between heater and chassis ground indicate either a leaky magnetron or hv diode.

One end of the diode was removed to check for leakage across the diode terminals. The resistance was about 3,700 ohms. The hv diode was replaced and another resistance check was made; now the resistance was 4,125 ohms. Perhaps, a new diode was leaky. The ground end was removed and the diode was normal. A resistance reading from heater to ground measured a little over 4 kilohms. Both the hv diode and magnetron were replaced. Of course, this is a very unusual case. You may find one component defective, but usually not both. The defective diode part number is 6000 000 0079 and 6000 000 00085 for the magnetron.

Norelco Model RR 7000—Dead—No heat

After replacing the 15 amp fuse, the fan rotated but there was no cooking. Poor line voltage was monitored at the primary winding of the hv transformer. High voltage was present with no current from the magnetron. Cooking was resumed with a new magnetron (part number 6000 000 00085).

Norelco Model MCS 7100—No heat—No cooking

No high voltage or current was found on the magnetron. The hv capacitor was discharged and resistance measurements were made. A resistance measurement of 71 ohms was found across the hv diode. The magnetron was suspected of a dead short. Of course, the low-ohm reading was still present with the heater leads removed. One lead of the hv diode was removed and the diode checked good. A quick resistance measurement of only 1.5 ohms was found across the hv capacitor. The 71 ohm reading across the diode to ground was actually the total reading of resistance of the hv power transformer winding and shorted capacitor (FIG. 12-16).

Norelco Model MCS 7100—Slow cooking

The complaint with this oven is that it took longer to cook food than usual. The low voltage across the power transformer primary winding was monitored with a vom. A current and hv meter monitored the magnetron. When the oven first came on, all voltage was fairly normal with about half the normal current reading.

After the oven was on for several minutes, the current would almost come up to where it should. Any time the high voltage is present with low or no current pulled by the magnetron, suspect a low-emission magne-

12-16 At first, a leaky magnetron was suspected when 71 ohms was measured across the hv diode. However, only 1.5 ohms was found across the hv capacitor. The 71 ohm reading was the result of the total transformer resistance plus the 1.5 ohms across the shorted hv capacitor.

tron tube. Replacement of the magnetron (number 6100 000 00029) solved the slow cooking problem.

Norelco Model MCS 8100—No heat—Fast count down

In this oven, no high voltage or power-line voltage was measured. No low voltage was found across the primary winding of the hv transformer. The controller seemed to be working except the count down appeared faster than usual. The vom was clipped across the primary winding to monitor the applied ac voltage.

Always discharge the hv capacitor before attempting to attach the voltmeter or take resistance measurements. A clip wire was connected across the triac connections, feeding the ac voltage directly to the primary winding (FIG. 12-17). The oven came on and worked as usual. Replacement of the leaky triac solved the hv and low-voltage problem.

Norelco Model MCS 8100—No cooking—No high voltage

When high voltage is not present, suspect problems in the hv circuits. Check the ac voltage across the primary winding of the hv transformer. Here, no voltage was measured. A quick glance at the magnetron tube showed an overheated thermal cutout. The wires to the thermal switch were burned and the cutout showed signs of excessive heat. A clip wire across the thermal switch applied ac to the hv circuits. The defective thermal switch was replaced with part number 8100 000 00098.

12-17 A clip wire was connected across the triac connection feeding the ac voltage directly to the primary winding. The oven then came on, indicating a defective triac assembly.

Sharp Model R6600—Dead—No cooking

Replaced blown fuse. High voltage normal. Replaced open magnetron FV-MZ001Y BKO.

Sharp Model R7600—Turntable rotates—No heat

Both the turntable and fan were rotating. No low voltage (120 Vac) on the primary winding of the hv transformer. Found a defective thermal cutout. The resistance of 5.1 ohms was measured across the switch terminals. No resistance should be measured across the thermal switch.

Sharp Model R7704A—Everything operates except cooking

Measured only 1.5 Vdc from heater terminal to chassis ground. Practically a dead short was measured from heater terminals to ground. Replaced shorted magnetron.

Sharp Model R7704—Will not cook

Installed a new magnetron tube FV-MXO 111 WRKO.

Sharp Model R7710—Not enough heat

The oven was checked out thoroughly at the shop. Cooking tests were perfect. When checked in the customer's home, only 105 Vac was measured at the ac outlet. Recommended electrician run a separate line from the switch box.

Sharp Model R7710—No heat—Defective diode

No high voltage was found after replacing the 15 amp fuse. Power-line voltage (120 Vac) was found at the primary winding of the hv transformer. Resistance measurements turned up a shorted diode RH-0Z0039W REO.

Sharp Model R7804—No voltage—No heat

In this model the thermal cutout was burned open, even the connecting wires showed signs of overheating. Ac voltage was now applied to the hv transformer, but no heat or cooking. Replaced defective magnetron FV-MZ0091-WRKO.

Sharp Model R7810—No heat—Burned cover

The customer had complained the cover inside the oven cavity was burning. The old cover was removed. Wipe out around the microwave waveguide outlet for excessive grease that may have collected there. A new magnetron FV-MZ0102 MBKO was installed to provide normal cooking.

Sharp Model R8310—No heat—Bad lead

Power-line voltage was applied to the primary winding of the hv transformer except there was no output voltage. Found poorly crimped leads from the transformer winding. Scraped around clip and crimped area. Applied rosin paste and soldered each transformer wire connection.

Sharp Model R9310A—No cooking—Bad leads

The complaint was poor heat and now no cooking at all. High voltage was present across the hv diode and no current measurement was found on the magnetron. Found excessive burned heater cables at the heater connections. Cleaned off connections and flat heater terminals. Soldered both connections.

Sharp Model R9310A—Lights up—No cooking

Found one bad heater connection on the magnetron. Cleaned off both connections and soldered each connection. Gave the oven the four-hour cooking test.

Tappan Model S6-1026-1—No heat

Found bad connections of crimped leads from the power transformer winding. Cleaned and applied solder paste. Resoldered each crimped transformer connection.

Thermador Model MC17—No heat—Voltage normal

Replaced defective magnetron L5261A.

OVEN GOES INTO COOKING CYCLE BUT THEN SHUTS DOWN

Norelco Model R 6008—Starts but no cooking

The oven quit operating after several minutes of cooking. All other functions were normal except no heat. Monitoring the voltage at the primary winding of the hv transformer indicated no voltage when the oven quit cooking. A new thermal cutout solved the shut-down problem. Clip a vom or 100 watt bulb across the thermal switch to see if it opens up (FIG. 12-18). The entire ac line voltage will be measured or the bulb will become bright when the switch opens.

12-18   Clip the ohmmeter leads or a 100 watt bulb across the thermal switch to see if it opens up. The entire line voltage will be measured here.

Norelco Model MCS 7100—Popped and quit

The complaint with this oven was that it was cooking beautifully then she heard a pop noise and the oven quit. No high voltage was measured. Resistance measurement of the power transformer was good. A leaky hv capacitor 6100 000 00005 popped and the oven ceased cooking.

Norelco Model 57500—Shuts off after several minutes

At first, a triac was suspected of shut down after no heat was noticed. A clip wire across the suspected triac indicated the triac was normal. The oven still quit after a few minutes of operation. Very little dc voltage was measured at the gate of the triac indicating a possibly defective control board. Before ordering a new control board, a new triac was installed, just in case it was shorting out the gate voltage. But a new control board brought the oven back to life.

Norelco Model MCS 8100—Just went dead after several minutes

Replaced the 15 amp fuse. No high voltage was measured on the magnetron. After several service checks, a leaky hv diode was replaced (6000 000 00079). Still the oven was inoperative. Replacing the control board 8100 000 00080 solved the "just went dead symptom."

Norelco Model MCS 8100—Ran one minute and quit

The ac voltage was monitored with a vom at the primary winding of the hv transformer. Found no voltage when the oven was not operating.

Everything else was working except no heat. Clipped a lead across the triac after discharging the hv capacitor. The oven continued to operate. The intermittent triac was replaced.

Sharp Model R6770—No cooking after five minutes

You would never suspect anything wrong with the oven if only a few minutes of cooking was required. Except after five minutes or longer the oven would not cook. Some overheated component was shutting the oven down.

The magnetron was too warm with the thermal switch opening up. When in the no cook operation the vari-motor was not rotating. After several hours of checking, the contact points of the temperature control assembly were opening up. Replacement of the assembly (QPWB F0002W REO) cured the shutdown problem.

Sharp Model R-7804—After a long period of cooking, quits

After 12 minutes of operation this microwave oven would shut down the cooking process. The fans and all other functions seemed to be operating. A lot of service time may be involved when attempting to locate the defective component with long cooking periods.

The oven would start to operate after it cooled down for a few minutes. Usually, heat shut down problems are caused by the magnetron tube. If possible, monitor the voltage on the primary winding of the hv transformer. Place a 100 watt bulb clipped across the thermal switch. Monitor the high voltage and current if a special meter is handy.

In this case the light bulb came on when the oven quit cooking. A new thermal switch did not solve the problem. Then a call was made to the factory service and a thermal kit was provided. A piece of insulation was placed behind the thermal switch, solving the long breakdown period.

ERRATIC OR INTERMITTENT COOKING

Litton Model 70/65.730—Normal high voltage

The hv and low-voltage circuits were monitored for a change in voltage. A current meter was tied to the leg of the magnetron. When intermittent, the current would drop to 0. The high voltage increased with no change in the low-voltage meter. Replaced defective magnetron.

Norelco Model MCS 6100—Intermittent operation

Found poor ac power in the home. Found improper line voltage. Requested electrician to install new outlet.

Norelco Model MCS 7100—Slow and erratic

An hv and current meter showed erratic current reading. The current would remain about half scale and then go up. Replaced magnetron 6100 000 00029.

Norelco Model MCS 7500—Erratic operation

An erratic low-voltage reading at the primary winding indicated a defective triac or control board. Shorting across the triac turned up a defective control board 7500 000 00/00.

Norelco Model MCS 8100—Opened door—Dead

Sometimes when the door was opened and then closed, the oven would become dead. Found an intermittent relay 8100 000 00011 (FIG. 12-19).

12-19 You should be able to hear any relay energize when voltage is applied. Although this relay is from an old oven, the operation is the same. Check the relay coil for continuity, and measure the voltage across the relay coil.

Norelco Model MCS 8100—Intermittent interlock

This oven may operate normally for several days, then become intermittent. Replaced erratic upper interlock switch assembly 6100 000 00055.

Quasar Model MQ6620 TW—Intermittent operation

Replaced intermittent latch switch 40-90344A79.

Samsung Model MW8600—Improper operation

The microwave oven did not cook as it should. Sometimes cooking took longer and at other times the oven lights seemed to blink. Even a closer inspection of the oven at the shop showed nothing unusual. Checking the home outlet and power-line voltage solved the improper operation. The line voltage measured 100 volts. Recommended a new outlet be installed by the local electrician.

Sanyo Model EM-8205—Erratic—Poor leads

Found poorly soldered or crimped on leads of the power transformer. Scraped back the enameled wire. Tinned and soldered each crimped on connection.

Sharp Model R5600—Erratic—Timer

This oven did not heat properly. Found the timer assembly was running slow. Sometimes the timer would not shut off at end of cooking cycle. Replaced defective timer QSWTE 0055WREO.

Sharp Model R6770—Erratic cooking

Replaced defective magnetron FV-MZ0078WRKO.

Sharp Model R7704A—Erratic heat

Incorrect time in cooking cycle. Replaced erratic timer QSWTE 0101WREO.

Sharp Model R6770—Erratic cooking—Erratic current

Monitored both high voltage and current of the magnetron. Replaced defective magnetron FV-MZ0078WRKO. In some ovens you can tell when the magnetron is heating up if heat is coming out of the vented blower area. However, this does not indicate the oven is cooking. Correct high voltage and current readings will determine if the magnetron is functioning properly.

Sharp Model R7710—Intermittent operation

Right side of the filament terminals were all black. Replaced defective clip and wire. Soldered both heater connections.

Sharp Model R7704—Intermittent—Magnetron

Monitored high voltage and current. Erratic current reading. Replaced erratic magnetron FV-MZ0011WRKO.

Sharp Model R7704—Dead, then intermittent

Replaced blown 15 amp fuse. Installed both monitor and interlock switch (QSW-M0100WREO and QSW-M00 64YBEO, respectively).

Sharp Model R9310A—Intermittent cooking

Burned heater wires. Replaced clips going over heater terminals. Cleaned off heater terminals. Soldered both connections.

Sharp Model R9310—Sometimes cooks

Found a defective interlock switch QSW-M0100WREO. Readjusted loose door.

Tappan Model 56-1026-1—Intermittent cooking

Sometimes the oven would cook and at other times not. Everything was normal except no heat. The filament of the magnetron tube was intermittent. Replaced magnetron. You can quickly tell if a magnetron is oscillating or cooking by setting a portable TV near the oven, while the oven is operating. A large noise band is generated across the TV screen on the lower TV channels (channels 2 to 5) when the magnetron is operating.

OVEN LIGHTS STAY ON

Norelco Model MCS 8100—Lights and oven stay on

The oven lights and cooking operation would stay on and not shut off when time was up. Replaced leaky triac assembly.

Magic Chef Model MW3117Z-SP—Light on

The light stayed on. Replaced latch switch. Readjusted door.

Sharp Model R9315—Oven light stays on

Replaced erratic secondary interlock switch QSW-M0096WREO.

KEEPS BLOWING FUSES

Amana Model RR-4D—Fuse blows when the door is opened

When opening the door, the fuse would open. Suspected defective monitor switch. Found metal lever at top of oven would not let the microswitch release. This safety switch would blow the fuse each time. Also replaced safety interlock switch.

Litton Model 465—Oven shuts off, blows fuse

When the oven would shut off sometimes the 15 amp fuse would open. Only after cooking for 10 minutes or longer would this condition occur. Replaced erratic triac.

Quasar Model MQ-6620TW—Blows fuse when plugged in

Each time the oven was plugged into the wall outlet, the fuse would blow. Found the varistor in the ac power line was arcing. This component is located in a clear plastic sleeve of the wire cable assembly (FIG. 12-20). Replaced.

12-20 A damaged varistor may cause the fuse to blow when the oven is plugged into the power source. The component may be located in a clear plastic sleeve of a Quasar Model MQ-6620TW.

Litton Model 370.00—Blown fuse, diode

The fuse would continually open in this microwave oven. A shorted hv diode was replaced.

Sharp Model R5600—Keeps blowing fuses

The interlock switch was hung up, so the monitor switch would keep blowing the fuse. Found food substance had run down inside the micro-switch and the plunger was held down. When the door was opened the plunger would not come up or release. Always replace the whole monitor and interlock assembly when one is found defective in a Sharp microwave oven.

Sharp Model R7710—Dead—Keeps blowing fuses

Replaced shorted magnetron tube FV-MZ0115MRKO.

Sharp Model R9315—Fuse opens—Hot operation

Found an interlock switch hang-up causing the monitor switch to open. Replaced both monitor and interlock switch assembly. The fuse would open when the oven would operate for over 12 minutes. Found the magnetron would short out. Here two different troubles caused the fuse to blow in the same oven.

Whirlpool Model REM 7400-1—Blows fuse when plugged in

As the ac cord was inserted into the power outlet, the fuse would open. Replaced frozen monitor switch. This switch is located in the middle of three separate switches.

HUM OR NOISY OPERATION

Amana Model RS 458P—Loud buzzing

The oven, when first turned on to cooking, made a loud buzzing and vibrating noise. With the cover off the noise was greater. Removed and replaced power transformer.

Hardwick Model M-250—Hum

Most noises caused in the microwave oven are the blower motor, radiating antenna, triac, and power transformer. Here the small controller transformer was causing the audible hum. Replaced the transformer.

K-Mart SKR7705A—Noisy turntable

When the turntable rotated you could hear a grating noise. Removed turntable assembly and replaced broken plastic coupling FCPL-001YWRKO.

Norelco Model MCS 8100—Noisy fan

Found loose fan mounting bolts on this one.

Norelco Model MCS 8100—Hum with closed oven

The oven would begin to hum when the door was closed. Removed and replaced leaky triac assembly 8100 000 00005.

Quasar Model MQ6600—Hot spot and noisy fan

The customer complained of a possible hot spot. The oven was normal. The fan was loose. Repaired and remounted the fan assembly.

Sharp Model R6770—Relay chatter

Sometimes the relay would begin to chatter when the oven switch was pushed. Erratic fan operation. Replaced erratic on/off switch.

Sharp Model R7704—Hum and buzz

When the cook button was pushed you could hear a loud hum. Removed the outside cover and now could hear a loud hum and buzz from the power transformer. Replaced noisy transformer RTN-00144WREO.

Sharp Model R7705A—Shock and noisy

Found a noisy timer. Replaced timer. Checked oven for poor grounding. Was normal in the shop. Found the oven was not grounded in the home.

Sharp Model R7804—Noisy turntable

Removed noisy turntable assembly. Lubricated and greased gear box assembly.

Sharp Model R8320—Fan noise at the top

In the convection oven a fan blade is located at the top of the oven. After removing several layers of metal ducts the fan blade was striking a metal bracket of the heating element. This fan is located in the round center circle of the heating element. Removed heating element and repaired mounting bracket. Replaced all components.

Sharp Model R9510—Constant buzzing

In the cook cycle the power transformer was buzzing very loudly (FIG. 12-21). Particles or iron vanes inside the transformer was making the noise. The noise would decrease when the load was taken off of it. Discharge the hv capacitor. Take one lead off of the heater connection of the magnetron. Replaced transformer RTRN-0160MKEO. These noisy transformers must be replaced as they are dipped and cannot be repaired.

12-21 You may find a noisy power transformer in the microwave oven, and this will need to be replaced. Here is a large power transformer in a Samsung RE-705IC model.

CANNOT TURN OVEN OFF

Norelco Model MCS 8100—Hums—Can't be turned off

In this oven, after running for 10 minutes the oven began to hum and would keep on operating when shut off. Replaced leaky triac 8100 000 00005.

Norelco Model MCS 8100—Fan fast operation—No shut-off

The fan motor seemed to run fast in this oven after shutting off. The oven would keep on operating in the cooking mode. Replaced leaky triac.

Whirlpool REM7400-1—Cannot turn oven off

Replaced defective on/off switch (FIG. 12-22).

12-22   Check the on/off switch when the oven cannot be turned off. These switches can stay on the on position. If the oven comes on when it is plugged into the power line, suspect a defective on/off switch triac or control board.

NO FAN OR BLOWER ROTATION

K-Mart Model SKR-7715—No fan rush

The fan was not rotating in this microwave oven. Discharged the hv capacitor. Measured the ac voltage across the motor leads. Found a bad fan connection when attaching the meter. Replaced and resoldered fan connection.

Litton Model 401—Slow fan blower

The oven fan would not come on for at least three to five minutes after the magnetron started up. The fan blade would not spin freely by hand. Spraying oil into the end and front bearings took care of the fan problem (FIG. 12-23).

12-23 Check the oven fan for dry or gummed up bearings when the magnetron is operating for a few minutes before the fan blower begins to operate.

Norelco Model MCS 8100—No fan rotation

Sometimes fan would start out slow and not rotate. Removed fan assembly. Cleaned up and lubricated fan bearings.

Norelco Model MCS 8100—Intermittent fan operation

Sometimes the fan would rotate and other times not. Repaired bad fan cable plug connection.

Sharp Model R7704A—Intermittent fan and oven

After five minutes of cooking the oven would shut down. When in the shut-down condition, the cooling fan was not rotating. Actually, when the fan stopped the magnetron overheated, kicking out the thermal switch. Found poor motor field connection. Repaired and soldered.

OVEN SHOCK AND LEAKAGE

K-Mart SKR-7705—Clicks and shocks

Checked ac leakage from metal chassis to ac power cord was normal. Customer suspected the oven of door leakage. Checked for leakage around the door and readjusted the door. Suspected static electricity since the kitchen floor area was carpeted.

Norelco Model MCS 7100—Intermittent operation—Shocks

The intermittent operation was caused by a two-wire extension cord. Low ac voltage was measured at the oven in the home. The oven was given a four-hour cook test in the shop. Installed a new three-prong outlet and grounded the oven.

Sharp Model R 5480—Customer afraid of shock

The oven had a burned waveguide cover and sparks were flying out the waveguide area. Replaced both waveguide covers and magnetron. The oven was checked for possible poor grounding. A thorough door leakage check was made.

Sharp Model R5600—Customer shocked by oven

The power cord and internal components were checked for ac leakage. Proper grounding of the oven in the home solved the possible shock hazard. Also checked for possible door leakage.

Sharp Model R7704—Customer suspects leakage

Radiation leakage was checked around the door and all vent openings. In fact, the oven was double checked for leakage, after the oven door was adjusted. Before door adjustment only 0.1mW/cm^2 leakage was noted at the top of the door.

Sharp Model R7705—Checked for shock and leakage

The oven and power cord were checked for ac leakage. The door and openings checked for radiation leakage. Requested customer properly ground the oven.

Sharp Model R7710—Sparks flying all over oven

The customer complained the inside of the oven had sparks flying all over. A new magnetron and waveguide cover were replaced. Magnetron part number FV-MZ0102 WRKO. Waveguide cover part number PC0VP0Z04WREO.

Sharp Model R7800—Sparks from oven

Replaced arcing magnetron tube FV-MZ00912 RKO.

TOO HOT TO HANDLE

K-Mart Model SKR-7705—Sparks in oven

Flashing and sparks flying in the oven. Replaced the defective magnetron and waveguide cover.

Montgomery Ward Model 67X175—Burning in oven

Here one plastic shelf was burning at one metal screw area. Excessive grease over the years had ran down behind the plastic shelves. Once the plastic started to burn, each time the oven was turned on, the burning would start up. Washed out all grease with soap and water. Removed shelves. Installed two new ones.

Norelco MCS 6100—Warped plastic cover

Popcorn was placed in a brown paper bag. The bag exploded and started a fire in the oven. The plastic shield and front door plastic was replaced. Some brown paper bags are made of recycled paper. If the bag is made from recycled paper, sometimes pieces of metal are found in the paper with the recycling process. The metal can cause the bag to start a fire which may end up destroying the microwave oven.

Norelco Model MCS 8100—Burns food

The owner complained that sometimes the oven would burn the food. Checked timing and control board. Cooking tests were normal in the shop. Found very high power-line voltage at the farm house (127 Vac). Referred the customer to the power company.

Sharp Model R5380—Oven smokes, won't cook

Replaced magnetron tube.

Sharp Model R5480—Something burning in oven

Removed burning waveguide cover. Washed out all grease. Installed a new cover PCOVP0039YBPO.

Sharp Model R6770—Too hot, burns food

This microwave oven was cooking too fast. A current meter revealed the magnetron was drawing excessive current. The magnetron would get extremely hot. Replaced defective magnetron FVMZ0078 WRKO (FIG. 12-24).

Sharp Model R7704A—Something smoking

After the 15 amp fuse was replaced, the oven began to smoke. Before the cord could be pulled, the fuse opened up once again. The suspected magnetron showed a leakage between heater and ground of 48.7 ohms with a digital vom (FIG. 12-25). The magnetron FV-MZ0091WRKO was replaced.

The oven was turned on and something appeared extremely hot. The hv capacitor was discharged. The secondary winding of the power transformer RTRN-KL44WREO was extremely hot. Undoubtedly, the shorted

12-24 In a Sharp Model R6770, the oven was burning food and food was cooking too fast. Insert a 10 ohm 10 watt resistor in series with the hv diode and measure for high current.

12-25 A leaky magnetron caused the power transformer to smoke in this Sharp R7704A oven. The resistance between the heater and ground was only 48.7 ohms. Both the magnetron and the transformer were replaced.

magnetron broke down the hv winding before the fuse was blown. Both the transformer and magnetron were replaced.

Sharp Model R-7404—Smokes and quits

Replaced 15 amp fuse. Replaced magnetron and waveguide cover.

Sharp Model R7650—Arcing in oven

Found egg shells went up through the open waveguide assembly. Excessive arcing from magnetron to ground. Cleaned up magnetron antenna area. Installed new metal gasket PGSK-0004YBEO.

Sharp Model R7800—Smoking in oven cavity

Replaced waveguide cover. Installed new magnetron. Original part number FV-MZ0091WRKO was subbed at the factory for a FV-MZ0 111WRKO.

Sharp Model R7844—Firing inside, now dead

Replaced 15 amp fuse. Replaced defective magnetron FY-MZ0111WRKO.

Sharp Model R9500—Sharp arcing

Found the fuse blown. Right away the fuse opened when a large arcing was heard. The magnetron tube was arcing internally. Replaced magnetron FV-MZ0091WRKO.

Sharp Model R9504—Burning food

Replaced defective magnetron FV-MZ0091WRKO.

Sharp Model R9750—Sparks, now dead

Replaced 15 amp fuse. Replaced waveguide cover and defective magnetron FV-MZ001WRKO.

CONTROL PANEL PROBLEMS

Goldstar Model ER-505M—Does not count down

When the start button was pushed the setting of time did not start to count down. Interlocks were normal. Suspected the control panel board (PCB). When the oven was fired up again it worked okay. When the PCB was pushed the next time the operation was intermittent. Repairing the loose connection at pin 10 solved the count down operation (FIG. 12-26).

Goldstar Model ER-410M—No cooking

The blower motor operated with no cooking. No line voltage was found across the primary winding of the hv transformer. Line voltage was found across the triac indicating a defective triac or no gate voltage. No gate voltage was measured at the G terminal of the triac. Replacing the control board solved the no cook problem.

Hardwick Model M270—Incorrect numbers

The numbers appeared incorrectly in the display tube. Replaced the control board assembly with microprocessor.

Norelco Model MCS 8100—Opened door—Dead

Replaced 15 amp fuse. Control board would not count down properly. Replaced control board 8100 000 00008.

12-26 A poor connection on pin 10 of the controller produced intermittent gate voltage for the triac in this Goldstar ER-505M oven.

Norelco Model MCS 8100—Intermittent operation

Monitored voltage at the primary winding of hv transformer. Suspected a defective triac when intermittent with no voltage applied to transformer. Shunted triac with a clip lead. The oven began to operate. Monitored voltage at gate terminal of triac with clip lead removed. Found gate voltage was very low when the oven was intermittent. Replaced control circuit board 8100 000 00008.

Norelco MCS 8100—Intermittent

Sometimes several push buttons in the control panel would engage and other times not. Ended up replacing front control panel assembly of controller unit.

Norelco Model MCS 8100—Behind "8" ball

The right-hand side number 8 was always shown behind any number. When the oven was turned on, you could see a faint number 8 in the background. A new control board was ordered and installed.

Sharp Model R9750—Time card error

When the time card was pushed in on this model, the cooking timer showed up as an error. Replaced entire card reader assembly DPNC-S0004PAZZ.

DOOR PROBLEMS

Hardwick Model M-240—Steam at door

In this model the steam seemed to flow into the door while cooking. The door seals were checked with the air vents. Improper venting and air flow allowed steam to enter the door area.

Litton Model 70-05—Dead—Leakage

Replaced 15 amp fuse. Checked cooking and leakage test. Found $0.25mW/cm^2$ leakage at top of door. Readjusted door and made another leakage test.

Norelco Model MCS 6100—Door open

The owner said the oven blew up. Found both door latches broke. The door would not catch or lock. Replaced both latch switches 5100 000 00055. Gave oven four-hour cooking test. Checked leakage after adjusting the door assembly.

Norelco Model MCS 7100—Door doesn't close

This door appeared warped or someone tried to adjust it. Removed door entirely and checked for warped door. Door assembly okay. Installed door and adjusted (FIG. 12-27). Checked for leakage.

12-27 Check the door hinges for proper adjustment. With this door, the two hex nuts are loosened, letting the door slide sideways and backwards. Tighten the nuts after the door is in place.

Norelco Model MCS 8100—Check door and shield

The customer complained of a warped shield 6100 000 00020. Ordered and replaced. Repaired and adjusted door for no play between door and base cabinet.

Norelco Model MCS 8100—Stuck door

Removed door. Repaired door guide assembly. Installed door assembly and adjusted for too much play. Made leakage and cook tests.

Norelco Model MCS 8100—Door will not shut

Replaced lost screw in the switch housing. The switch assembly was down so the hook would not lock in. Adjusted door. Made leakage test.

Norelco Model MCS 8100—Binding door

Readjusted door latch assembly. Readjusted door assembly. Checked other interlock switch assemblies. Made cook and leakage tests.

Sharp Model R3700—Loose door latch

Found a loose rivet and the door assembly had too much play. Repaired door latch assembly with a new pop rivet.

Sharp Model RS480—Door won't open

The door was stuck on this microwave oven. Pushing down on the latch would not release the door. After removing the back cover and discharging the hv capacitor, one of the door latch assemblies helped to release the door. Replaced broken door latch assembly.

Sharp Model R6770—Intermittent operation

Sometimes the oven would operate and at other times quit. Found a broken door spring in the door assembly. Replaced spring and readjusted door. Checked for leakage (FIG. 12-28).

Sharp Model R7600—Blows fuse

Every once in a while the fuse would blow for no known reason, until the operator said the oven quit when the door was opened. Checked door for correct closing. Replaced left-side interlock assembly and also readjusted door. Checked cooking and leakage tests.

Sharp Model R7600—Must raise up door to operate

The owner had to raise up on the door to make the oven operate. Replaced door interlock monitor switch assembly.

12-28 Check for a broken door spring behind the plastic latch hook. Replace the spring and readjust the door, then check for leakage.

Sharp Model R7650—Oven light—No heat

Fuse was good with oven lights on. All interlocks were checked and found okay. Located broken spring in the door latch assembly. Replaced, checked leakage, and gave a cooking test.

Sharp Model R7650—Intermittent—Now dead

Replaced 15 amp fuse. Realigned the front door. Checked for door leakage.

Sharp Model R7650—Door loose

The oven would begin to operate when one pushed against the door. A new interlock monitor switch was installed. To keep the switch assembly in place so the latch head will contact the microswitch, slide the assembly tight against the front piece. Hold the unit in place by placing a thin screwdriver blade in the mounting slot, then push the assembly forward as you tighten the nuts holding the switch assembly.

Sharp Model R7710—Intermittent cooking

When in the intermittent state the whole oven would shut down. Readjusted door and interlock switches.

Sharp Model R7804—No heat

No heat and no cooking were the oven symptoms. When loaded oven for a water cook test and when closing the door, sometimes the oven may come on. Readjusted door and checked interlock assemblies.

Sharp Model R9200—Poor door closing

Adjusted door. Replaced damaged control panel assembly HPNLC1017C-CZZ. Checked for leakage.

Sharp Model R9310A—Broken door

Even the glass was broken out of the door. Undoubtedly, something was thrown at the microwave oven. Replaced entire door assembly DD0RF0129WRKO. Adjusted door and made leakage tests.

Sharp Model 9310—Loose door

Found too much play between door and oven base plate. Sometimes oven would quit operating when door was pulled outward. Adjusted door and interlock switch assemblies. Made cooking and leakage tests.

LIGHTNING DAMAGE

Norelco Model MCS 8100—Control board

Had to replace the entire control board in this oven, which had lightning damage. Check the oven for a damaged thermistor. Sometimes only the thermistor and maybe some wiring needs to be replaced. If any doubts exist about the control board operation, replace it. Control board part number 6000 000 00010.

Sharp Model R6460—Lightning damage—Dead

Replaced 15 amp fuse. Replaced blown apart interlock switch QSW-M0039YBEO.

Sharp Model R8310—Temperature control board

Replaced blown fuse. Oven still dead. Checked and found oven temperature control board was damaged. Removed and replaced entire assembly 2KBBK0106WRVA (FIG. 12-29).

MISCELLANEOUS OVEN PROBLEMS

Hardwick Model MM-270—Longer cooking

The complaint of this oven was it appeared to take a little longer than normal to cook food. Perhaps the magnetron was getting weak, but found the radiating antenna was not rotating. The antenna belt was off.

12-29 Notice that the wiring was damaged and there are black markings on the metal chassis of the temperature-control assembly. If lightning strikes the temperature control or control circuit board, always replace it.

Norelco Model MCS 6100—Warped cover

Top waveguide cover warped and part hanging down. Replaced cover 6100 000 00020. Checked cooking and leakage tests.

Norelco Model MCS 8100—Keeps running

Replace shorted triac assembly 8100 000 00005.

Norelco Model S-7500—Power outage

After a lightning and wind storm, replaced control board assembly. Varistor and wiring on control board damaged (FIG. 12-30).

Norelco Model MCS 8100—Oven won't shut off after 15 minutes

This oven operated okay on small cooking warm-up under 15 minutes. When food was cooked over that time, the oven would not shut off and the power plug must be removed. Replace leaky triac assembly.

Samsung Model 70STC—Firing in door

Found too loose screws in the door. Checked door screws for leakage and arcing. Sometimes arcing will occur around metal screws inside the oven cavity.

12-30 Check the control board assembly after lightning or power outage has occurred. Here, a small varistor was damaged on the control board assembly.

Sharp Model R7650—Flickering lights

When the door was opened the oven lights would begin to flicker. The oven lights were normal with the door closed. Replaced the secondary interlock switch.

Sharp Model R7750—"Oven headache"

The customer complained of a headache when the microwave was operating (really afraid of radiation leakage). Brought oven to the shop. Checked oven for shocks, leakage, and proper cooking. Found nothing wrong. Returned oven.

Sharp Model R7710—Slow cooking

The customer complained of dead spots and too much time for the cooking process. Oven was okay on several water cooking tests. Checked oven in the home. Had only 112 Vac on extension cord when oven was operating. Suggested electrician install power outlet from the fuse box.

Sharp Model R7810—Temperature probe shuts off

After cooking two minutes with the temperature probe, the cooking probe shut off. Oven works okay without probe. The dealer exchanged probes, the same trouble occurred. The oven would shut off after two or three minutes of cooking with the probe. Replaced entire temperature probe electronic assembly FPWBF-0007WRKO.

Sharp Model R8010—No convection cooking

Microwave cooking normal. No convection cooking. In this oven the heating element is on one side and consists of a round element. The select switch was turned to convection cooking. The fan motor turned with no heat. Removed cover and discharged capacitor. Checked resistance across heating element (FIG. 12-31). Resistance should be under 10 ohms. No continuity measured. Ordered and replaced heating element.

12-31   In a convection oven, check for complete resistance across the heating element with the vom. Also, inspect the heating element terminals for burned or poor connections.

Sharp Model R8310—Poor convection cooking

The customer complained of poor heat from the heating element. Also a hot smell was noted. The microwave oven cooking was normal. Found fan blade was not rotating. Replaced broken belt going from magnetron fan motor to heating element. The heating element fan is located right in the center of the heating element blowing heat down from top of oven.

Sharp Model R9105—Moisture—Afraid of leakage

The customer complained of moisture fogging the glass at the bottom of the door. Afraid the oven was leaking radiation. Moisture can build up inside any oven depending upon what is being cooked. Check all vents and openings. Remove any dirt, dust, and lint from these openings. Check for enough space around the oven. The oven may be enclosed where the exhaust cannot properly take away the moisture. Check for door leakage.

Sharp Model R9210—Dead—Power voltage

After a storm and power outage, the microwave oven was dead. Replaced the 15 amp fuse. The fuse would keep blowing. Replaced burned line varistor with a GE-750 line protector. Repaired two wires burned on the circuit board. Checked cooking and gave leakage tests.

Chapter **13**

Important
Dos and Don'ts

FOR SAFETY

Don't operate oven cavity empty. Do use at least a liter of water or a paper cup full of water in the oven cavity when making tests.

Don't overload the line voltage circuits. This can cause the oven to operate erratically or intermittently in the home. Do check the line voltage and the oven should be operated from a separate ac outlet.

Don't let the owner use a light extension cord for oven operation. Do suggest a separate outlet and always check for proper grounding of the microwave oven.

Don't operate or allow the oven to be operated with the door open. Do warn the owner of possible injury when the oven operates with the door open. Simply replace the defective monitor switch.

Do check the following components before servicing the oven: interlock operation, proper door closing, seal and choke surfaces, loose or damaged hinges, and evidence of cabinet damage (FIG. 13-1). Don't forget to replace all damaged oven door components.

Don't under any circumstance place your hands in the oven while the oven is operating (FIG. 13-2). Do pull the power plug before testing or removing the suspected component. You may receive a terrible burn or shock that can be fatal.

Don't forget to remove the glass turntable tray or cooking test equipment from the oven before doing actual repair. Do remove everything from the oven cavity. Replacing a broken turntable tray may cut the repair bill in half.

Don't work around the microwave oven when the owner or several people are hovered around asking questions. Do keep a cool head and always think what should be done for quick repair and safety.

13-1   Check the oven door for evidence of cabinet damage or when the door has too much play.

13-2   Don't under any circumstances place your hands in the oven while the oven is operating. Always pull the power cord before testing or removing component.

Do keep your hands away from the ICs and critical components when replacing an electronic control board. Handle the board by the edges, keeping your body at ground potential with the metal oven cabinet (FIG. 13-3). Don't just rip off the insulated protection foil and slap the board in place.

Don't ever remove a clip over the contacts of a defective interlock switch and leave it that way as a repair. Do replace all interlock switches with the original replacement part when available.

Do several leakage tests when the customer is afraid and complains of possible oven leakage. Don't forget to report a dangerous leaky oven to the manufacturer if more than 2mW/cm^2 leakage is found in the oven. Warn the owner not to use the oven until properly serviced.

13-3 Do handle the control board around the edges, keeping your body at ground potential. Wear a wrist ground strap, and keep your hands off the critical IC components.

Do check the magnetron, waveguide assembly, and microwave generating components for possible damage and leakage. Don't let the oven go out without proper leakage and cooking tests.

FOR PROPER OVEN REPAIRS

Don't forget to discharge the high-voltage (hv) capacitor after pulling the ac cord. Do pull the plug and discharge the two capacitor terminals with a metal screwdriver (FIG. 13-4). Make sure the capacitor terminals are discharged, not just from one side to the metal case.

Don't forget each time to pull the power cord and discharge the hv capacitor when taking continuity and resistance measurements with the vom. Do double check the discharging of the hv capacitor. Look for an arc or spark when the two terminals are shorted.

Do use alligator clip leads when connecting test instruments to the various components. Don't hold the test probes in your hand when taking actual voltage measurements (FIG. 13-5). Insulate the test instrument with a book or rug for possible ground with the test instrument set on the metal cabinet.

Don't ever attempt to use an ordinary vom in making hv measurements. Do use a correct hv probe or hv test instrument.

Don't stick any metal tools such as screwdrivers or nut drivers in the oven while it is operating. Do be careful when the oven is cooking. Only use a piece of wood or plastic-type probe tools.

13-4 Don't forget to discharge the hv capacitor. Make sure the two capacitor terminals are discharged.

13-5 Don't hold the test probe in your hands when taking actual voltage measurements. Instead use test clip leads.

Don't leave the meat probe on the metal oven cavity with the oven operating. Do remove the probe and cable when checking or making tests on the oven.

Don't forget to open and close the oven door several times to check for a possible defective interlock switch. Do make sure the oven door fits snug with no play.

Do use a 100 watt lamp bulb with connecting cable clips across the 15 amp fuse terminal when the oven keeps blowing fuses. Don't forget the lamp is only an indicating device and will be very bright when a shorted component is still in the oven.

Don't forget to mark down where each wire goes when removing a defective component (FIG. 13-6). Always mark the wiring leads in the service manual with all components having more than two connecting wires. Do be careful. Correct component replacement saves time and money.

Don't substitute oven parts if they do not have the same rating or physical size. Do remember the hv capacitor and diode, interlock switch, and magnetron tubes may be interchanged with the same or other manufacturers, providing they are the same part number and physical size.

Do follow the manufacturer's instructions in replacing the control board. Don't forget to read the instructions for checking off the various components before replacing the control board.

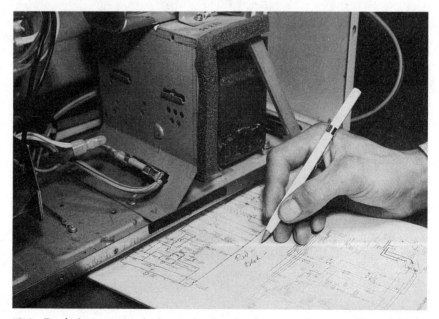

13-6 Don't forget to mark down where each wire goes when removing a defective component. These color-coded leads can be marked right on the service manual.

13-7 Don't forget to clean out the air vents and dust the top vent areas.

Don't forget to replace all mounting screws in components or the back cover of the metal cabinet. Do check the front of the oven cabinet for loose or outside tab mounts before the cover is locked in place.

Don't forget to run a cooking test on the oven after all repairs are made. Do keep the oven operating for a two hour test on the service bench. Now is the time to take that final leakage test.

Don't forget to make a leakage test after each repair. Do make a careful leakage test around the door area. After replacing the magnetron, always make a leakage test around the waveguide area.

Don't forget to clean out the air vents and intake areas (FIG. 13-7). Do brush out all vents and clean out all pieces of food collected at the top vent area.

Do clean up the outside cabinet and controls with window spray or cleaning detergent before returning the oven to the customer. Don't forget to take out the cooking test equipment. You will need it for the next oven repair.

Don't forget to remove your wristwatch while working around the oven. Do take off the watch when replacing the magnetron tube.

Chapter **14**

Where and how to obtain parts

After locating a defective component, securing the part may take a little longer. Although some oven components are interchangeable, they should be purchased from the correct oven manufacturer (FIG. 14-1). Most oven manufacturers issue the parts or they have distribution centers around the country who can get them for you. Go directly to local chain stores, such as Sears and Wards to order their oven components.

Many microwave oven manufacturers have service depots and personnel located in various states for warranty repair service. Just give them a call. You may find a list of warranty stations enclosed in the oven operation literature. Call or write to other manufacturers listed below to obtain service manuals and oven parts.

AMANA REFRIGERATION, INC.
Amana, IA 52204
(319) 622-5511

CALORIC CORP.
Topton, PA 19562
(215) 682-4211

CROSLEY CORP.
P.O. Box 1959
Winston-Salem, NC 27102
(919) 761-1212

I4-I Although some oven components are interchangeable, they should be purchased from the oven's manufacturer, when possible. Some magnetrons, timers, relays, and control boards cannot be interchanged.

EMERSON RADIO
One Emerson Lane
North Bergen, NJ 07047

FRIGIDAIRE DIV. WC1
300 Phillips Road
Columbus, OH 43228
(614) 272-4100

GENERAL ELECTRIC CO.
Appliance Park
Louisville, KY 40221
(502) 452-4311

GOLDSTAR ELECTRONIC INC.
1050 Wall Street West
Lindhurst, NJ 07071
(201) 480-8870

HARDWICK STOVE CO.
740 King Edward Avenue
Cleveland, TN 37311
(615) 478-4610

HITACHI SALES CORP. OF AMERICA
401 West Artesia Boulevard
Compton, CA 90220
(213) 537-8383

HOTPOINT
General Electric Co.
Appliance Park
Louisville, KY 40225
(502) 452-4311

J. C. PENNEY CO.
(Check at local store)

JENN-AIR CORP.
3025 Shadeland Avenue
Indianapolis, IN 46226
(317) 545-2271

KITCHEN AID
701 Main Street
St. Joseph, MI 64501
(616) 982-4500

LITTON MICROWAVE COOKING PRODUCTS
4450 Mendenhall Road South
Memphis, TN 38101
(901) 366-3000

MAGIC CHEF
740 King Edward Avenue
Cleveland, TN 37311
(615) 472-3371

MAYTAG
One Dependability Square
Newton, IA 50208
(515) 792-7000

MODERN MAID
Topton, PA 19562
(215) 682-4211

MONTGOMERY WARDS
(Check with local stores)

NORELCO AMERICAN PHILIPS
High Ridge Park
Stanford, CT 06903
(203) 329-5700

PANASONIC CO.
One Panasonic Way
Secaucus, NJ 07094
(201) 348-7185

QUASAR
1325 Pratt Boulevard
Elk Grove Village, IL 60007
(312) 228-6366

ROPER SALES CO.
1507 Broomtown Road
Lafayette, GA 30728
(404) 638-5100

ROYAL CHEF
Gary & Dudley
2300 Clinton Road
Nashville, TN 37209

SAMPO OF AMERICA
5550 Peachstreet Ind. Boulevard
Norcross, GA 30728
(404) 449-6220

SAMSUNG ELECTRONICS
301 Mayhill Street
Saddlebrook, NJ 07662
(201) 587-9600

SANYO-FISHER, INC.
21350 Lassen Street
Chatsworth, CA 91311
(818) 996-7322

SHARP ELECTRONICS CORP.
Sharp Plaza
Mahwah, NJ 07430
(201) 265-5600

TAPPAN APPLIANCE DIVISION
300 Phillips Road
Columbus, OH 43228

TATUNG OF AMERICA
2850 El Presido Street
Long Beach, CA 90810
(213) 637-2105

THERMADOR/WASTE KING
5119 District Boulevard
Los Angeles, CA 90040
(213) 562-1133

TOSHIBA AMERICA, INC.
Home Appliance Division
82 Otawa Road
Wayne, NJ 07470
(201) 628-8000

WHIRLPOOL CORP.
2000 M-63 North
Benton Harbor, MI 49022
(616) 926-5000

WHITE-WESTINGHOUSE
Div. WC1
300 Phillips Road
Columbus, OH 43228
(614) 272-4100

SERVICE LITERATURE

Manufacturers' service literature is one of the most important pieces of information in servicing microwave ovens. Besides several circuit diagrams and parts lists, you may find trouble charts on how to service a certain model. Invaluable tips and charts are found in the oven service literature (FIG. 14-2).

Very few manufacturers include voltage or current measurements on the circuit diagram. When any given oven has been repaired and is operating correctly, it's wise to take voltage and current measurements. Just place the measurement right on the circuit diagram for future reference. Remember, high voltage (hv) and current readings must be taken with an instrument manufactured especially for microwave oven repair.

When unusual or difficult problems are found in the oven, they should be marked on the schematic for future reference. Often these same problems crop up and it's difficult to remember all of them (FIG. 14-3). These difficult problems may be compared with problems with other ovens, saving valuable service time. If the service manual is not handy, one may be ordered directly from the manufacturer.

WARRANTY PARTS

Most oven manufacturers' components are warranted for one or two years. The magnetron may be guaranteed from five to seven years. Periodically a warranty service shop is issued the warranty description for each oven. You may find the warranty period listed in the owner's manual.

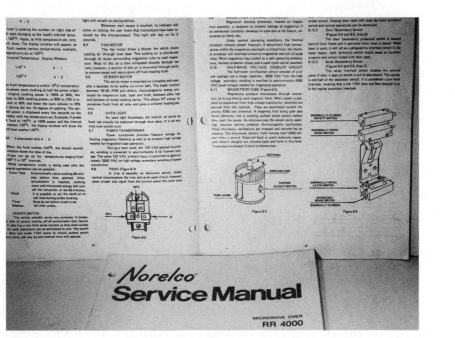

14-2 Invaluable tips and charts are found in the oven service literature. Check the trouble charts for possible oven symptoms (courtesy of Norelco Corp.).

14-3 Mark down the most difficult problems in the service manual, so they will be there the next time you need them.

Oven parts found to be in the warranty period must be treated a little differently than those out of warranty.

With some manufacturers, only the larger components are returned in the warranty period. Others request only the magnetron tube be returned. For instance, Sharp Electronics Corporation requests that the warranty station return all components except small items, such as switches, fuses, and hv diodes. Norelco may request the return of warranty components such as control boards and magnetrons. Each oven manufacturer may have a different warranty replacement procedure. All warranty claims should be followed correctly to receive credit and pay for services rendered.

HOW TO ORDER PARTS

When ordering a defective component, always include the part number, model number of the oven, and component description. The defective part may be located in the parts list of the service manual. Some ovens have both a physical and part drawing number for locating each component. Also, listing the reference number, if handy, may help secure the correct component. Simply find the part in the layout drawing, then check the number in the part number.

You must include all numbers or letters in the part number. Often, the oven part number is rather long. For instance, in a Sharp microwave oven, there are a total of 13 numbers and letters in the part number. The waveguide cover part number for a Sharp R9330 model is PC0VP0204WRE0. A turntable coupling assembly for the same oven is FCPL-0018WRK0. Notice the dash mark is included in the total of 13 letters and numbers. While in a Norelco oven, the part number has a total of 12 numbers. The first four numbers are the model number of the oven. A controller board in a MSC 8100 Norelco oven has a part number 8100 000 00008. You may find the part number may begin with a 6100 or 7100 number which shows the component is common to both ovens (FIG. 14-4).

In case a service manual is not handy, the component may be ordered by giving the model number of the oven and number marked on the part. You may find a magnetron tube number stamped directly on the body is the same as the part number. It's best to order the oven service manual and then order the component from the parts list. You are ensured of obtaining the correct component with this method.

RETURNING CRITICAL COMPONENTS

All critical parts such as the magnetron, transformer, and control board should be sent back in the original cartons. Often, these components are packed in two different boxes. Special magnetron boxes have additional packing to prevent breakage to the glass portion of the magnetron. Inspect the vent area of the magnetron for possible damage before installing the new tube.

14-4 Usually the numbers found on the side of the component are not the part number, so look it up in the service manual.

Controller or circuit boards are packed in a special nonstatic sleeve and are taped up (FIG. 14-5). Replace the circuit board in the same enclosure, insert in the part box, and then place the box inside a larger one for protection. Never send a defective power transformer and a circuit board back in the very same carton. The heavy transformer may break or otherwise damage the circuit board. You should have no problems if the defective component is returned in its same shipping carton.

Before sealing the final box, enclose the warranty service report. Most oven manufacturers require that the service dates be included with the defective component. In fact, when several different warranty components are returned, tape the service form to each defective part. Since many different oven manufacturers have their own service form, it's best to follow their warranty part procedures.

SUBSTITUTION OF PARTS

Although the manufacturer requests that both the oven technician and the repairman use exact part numbers, there are a few components that can be purchased from electronic distributors, warehouses, or mail-order firms. The biggest problem in substituting parts is the physical size (FIG. 14-6). Replacement components should have the same operating voltage or higher with correct wattage.

Always replace with original part numbers when possible. Sometimes the manufacturer may be out of business or critical parts may be

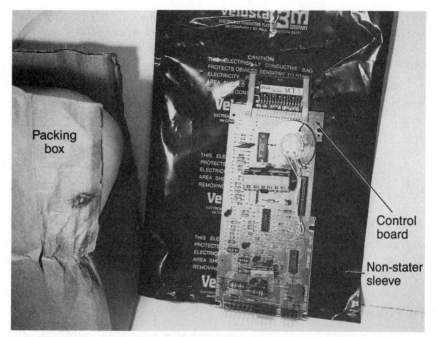

Packing
box

Control
board

Non-stater
sleeve

14-5 Controller or circuit boards are packed in a special nonstatic sleeve. Then pack the complete package inside another one to prevent board breakage.

14-6 Make sure the magnetron replacement is the same size and wattage, and that the mounting holes fit and the replacement has the same direction of air flow from the fan motor.

tied up for sometime, resulting in parts substitution. It's possible that warranty replacement parts pile up in the dealers stock, and they should be used up as soon as possible. Here are a few components that may be substituted:

Fuses

The microwave oven employs a ceramic-type fuse about $1^1/4 \times {}^1/4$ inch in size. These fuses may come in 10, 15, 20, 25, and 30 amp. The common ceramic fuses are 15 and 20 amp. These fuses can be purchased almost anyplace where ovens are sold. Replace with exact amperage as the original.

Interlocks

A lot of the interlock switches can be exchanged from one oven to another. Make sure the switch is the same size, closed or open, or two switches in one and a single switch.

Hv capacitor

Replace the capacitor with the exact working voltage or higher, and the same capacitance or higher. These capacitors are easily interchanged between ovens.

Hv rectifier

Microwave oven diodes can be subbed with ECG 541 and 548 replacements. RCA 9306 and 9305 are 1,000 and 500 MA forward current types. Pick these up at wholesale parts houses, distributors, and mail-order firms.

Magnetrons

Sometimes magnetron tubes can be interchanged, since several manufacturers make ovens for many different firms. Make sure the tube will fit physically and has correct wattage and fan blower application. Make sure the hv and heater terminals are in the same direction. Match up the correct mounting and direction of air flow from fan blower. Remember, some of these magnetrons have a fan blower mounted on the bottom side.

Do not install a magnetron from a small 250 – 400 watt oven into a 600 – 700 watt oven; nor should you install a 600 – 700 watt magnetron tube in a 400 watt oven. Check the oven for an external hv rectifier. If the diode is inside the tube cavity, you cannot replace it with one without. Most magnetron replacements do not contain an internal rectifier. If the magnetron replacement has the same wattage, physical size, and correct direction of air flow, no doubt it will work satisfactorily.

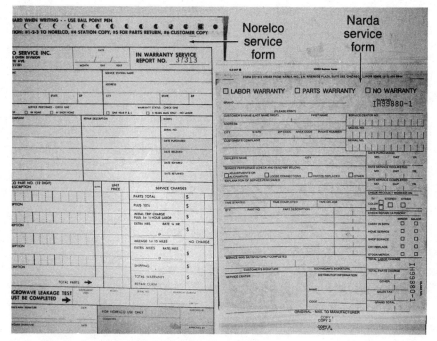

14-7 Fill out all spaces in the service form. Some firms have their own forms, while others may use the Narda universal service form.

SERVICE FORMS

Although some oven manufacturers will issue their own warranty forms for service and parts reports, many manufacturers prefer the Narda service form number 317-515. The Narda service report may be used for television, radio, stereo, and oven service reports. This service form consists of the customer's name, address, and where the oven was purchased (FIG. 14-7). Don't forget to include the model and serial number of the oven (also, the date purchased, received, and repaired). Some manufacturers request a microwave leakage test be shown. This leakage reading should be marked on all microwave oven reports for the service technician and manufacturer's protection. All Narda warranty business forms may be obtained from Narda, Inc., 2N Riverside Plaza, Suite 222, Chicago, IL 60606. Telephone (312) 454-0944.

Appendix

Selected
Circuit Diagrams

A-I   The wiring schematic of a Norelco RR7700 microwave oven (courtesy of Norelco Corp.).

PARTS LIST

Item No.	N.S.I. Part Number	Description	
1	6000-000-00001	Wrapper, Back and Bottom	805T015F15
2	6000-000-00003	Wrapper, Top and Sides (Walnut)	809T011F04
3	7500-000-00013	Angle, Air Baffle	008T378F05
4	7500-000-00014	Angle, Probe Receptacle	008T435F01
5	6000-000-00006	Baffle, External Vent	024T123F03
6	6000-000-00007	Motor, Stirrer	463T070P07
*	6000-000-00008	Screw, Stirrer Motor (10Ax3/8" Ph. Tr.Hd) (2)	636T091P01
7	6000-000-00009	Fan, Stirrer	239T028F01
8	6000-000-00010	Shaft, Stirrer	652T005P01
*	6000-000-00011	Screw, Stirrer Shaft to Fan (6-32x3/8" Ph. Pan Hd) (2)	636T095P31
*	6000-000-00012	Nut, Stirrer Shaft to Fan (6-32 W/Ex. T Washer) (2)	479T069P01

* Not illustrated

A-2 The basic body layout of a Norelco S7500 microwave oven with important oven components listed (courtesy of Norelco Corp.).

Item No.	N.S.I. Part Number	Description	
9	6000-000-00013	Fastener, Shaft Retainer	241T017P151
10	6000-000-00014	Grommet, Stirrer Pulley (2)	287T026P01
11	6000-000-00015	Belt, Stirrer Drive	044T002P02
12	6000-000-00016	Bushing, Stirrer Shaft	102T027P01
13	6000-000-00017	Washer, Stirrer Shaft	789T022P03
14	6000-000-00018	Channel, Air Duct	136T258F04
15	6000-000-00020	Bushing, Strain Relief	102T024P03
16	6000-000-00021	Cord, Power Supply	171T019P04
17	7500-000-00015	Stirrer Cover	174T165P04
18	6000-000-00023	Cover, Light Access	174T169F01
19	6000-000-00024	Duct Bottom	211T032F05
20	6000-000-00025	Fastener, Stand Off	241T017P153
21	6000-000-00026	Fastener, Purse Lock (Ground Wires)	241T017P154
22	N/A	Front Main Wave Guide	N/A
23	6000-000-00027	Tray, Oven (Glass)	736T005P01
24	6000-000-00028	Glide, Tray (4)	280T046P01
25	6000-000-00029	Screw, Glide Mtg. (10AB x 5/8" Ph. Fillister) (8)	636T091P19
26	6000-000-00030	Socket, Light	662T027P02
27	6000-000-00031	Shield, Oven Light	655T302P02
28	6000-000-00032	Light Bulb	374T018P01
29	6000-000-00033	Grommet, Door Strike	287T003P08
30	6000-000-00034	Cover, Microwave Seal (2)	174T166P02
31	6000-000-00035	Seal, Microwave (2)	643T008P09
32	6000-000-00036	Spacer, Foot (4)	665T132P02
*	6000-000-00037	Screw, Foot and Bottom Trim Mtg. (8Ax3/4" Ph.Tr.Hd) (4)	636T090P10
33	6000-000-00038	Trim, Bottom	740T385P05
*	6000-000-00039	Screw, Bottom Trim Mtg. (8A x ½" Ph. Tr. Hd)	636T069P07
34	6000-000-00040	Trim, Top	740T386P05
35	6000-000-00041	End Cap (4)	123T079P01
36	6000-000-00042	Medallion, Door Strike	445T004P01
37	7500-000-00016	Fastener, Wrapper (3)	241T017P152
38	6000-000-00044	Trim, Front Main	740T392P02
39	6000-000-00045	Bushing, Wire (2)	102T028P01
40	7500-000-00017	Triac, Heat Sink Assembly	742T001P02
41	7500-000-00018	Fastener, Triac (3)	241T017P198
*	7500-000-00019	Screw, Wrapper (8A x ½" W/Int. Washer) (23)	636T090P46
42	7500-000-00020	Probe, Roast Control	563T008P01
43	7500-000-00021	Receptacle, Probe	597T029P01
44	7500-000-00022	Nut, Receptacle Mtg.	479T082P01
45	7500-000-00023	Insulator	331T117P01
46	7500-000-00024	Insulator, Front Frame (Cap to Touch Panel)	331T118P01
*	7500-000-00025	Kit, Touch-up Paint (Almond)	508T020P02
47	7500-000-00026	Cover, Corner (2)	174T213P01
48	7500-000-00027	Spacer, Eyelet (4)	665T155P01
*	7500-000-00028	Grounding Washer, Wave Guide (Int. Tooth) (4)	789T053P30
*	7500-000-00029	Screw (10A x ½" Ph.Tr.Hd.) Back Wrapper	636T091P02
*	7500-000-00030	Screw (8A x ½" W/Int. Washer) Wrapper	636T090P46

* Not illustrated

A-2 Continued

NN-8507/NN-8807/NN-8907 APH

NOTE: Door is closed.
Unit is not operated.
⏚ Ground
⏚ Chassis

IMPORTANT SAFETY NOTE: THE SHADED AREA ON THIS SCHEMATIC DIAGRAM INCORPORATES SPECIAL FEATURES IMPORTANT FOR PROTECTION FROM MICROWAVE RADIATION, FIRE AND ELECTRICAL SHOCK HAZARDS. WHEN SERVICING, IT IS ESSENTIAL THAT ONLY MANUFACTURER'S SPECIFIED PARTS BE USED FOR THE CRITICAL COMPONENTS IN THE SHADED AREAS OF THE SCHEMATIC DIAGRAM.

A-3 The schematic diagram of the Panasonic NN-8507/NN-8807/NN-8907 APH microwave ovens (courtesy of Matsushita Electric Corporation of America).

A-4 The wiring diagram of the Panasonic microwave ovens NN-8507/NN-8807/NN-8907CPH (courtesy of Matsushita Electric Corporation of America).

EXPLODED VIEW

(S-425/426 **APH, CPH, AQH, CQH**)

A-5 The exploded view of parts in the Panasonic Models NN-9807, NN-9507, MQ8897BW, MQ8897BW, and MQ8877BW (courtesy of Matsushita Electric Corporation of America).

A-6 The DPC of Panasonic's NN-9507, NN-9507P, and NN-9807 models (courtesy of Matsushita Electric Corporation of America).

G   H   I   J   K   L   M

Glossary

ac voltage Alternating current.

anode The positive electrode in the magnetron receives the electrons. In this case, the outer shell of the magnetron is the anode and is always grounded.

bell or buzzer An electronic or mechanical sounding device to indicate that the cooking process is finished. The bell may have a mechanical striker on the timer assembly. The electronic buzzer or speaker indicates a loud tone when the oven cooking cycle is over.

blower motor The blower motor keeps the magnetron cool. The blower motor circulates air and helps remove moisture from the oven cavity.

capacitance The electric size of a capacitor. Sometimes called capacity. The basic unit of capacitance is the farad. A capacitor permits the storage of electricity between two insulated conductors.

capacitor A capacitor stores electric energy, blocks the flow of direct current, and permits the flow of alternating current. The large capacitors found in the microwave ovens are in a metal container. Most hv capacitors in the oven are oil-filled types.

cathode The cathode element provides the source of electrons. In the magnetron tube the cathode is centered within the anode cavity and has a highly negative voltage potential.

choke An inductance used in a circuit to present high impedance to frequencies without limiting the flow of direct current. In the microwave oven, a groove or shaped area to reflect guided waves within a limited frequency range.

circuit A path in which electrical current can flow.

control board An electronic board used to control the time and operation of the oven cooking process. The controller or control board may be replaced separately or repaired.

current   The flow of electricity. The current found in the magnetron circuit is measured in milliamperes.

defrost   To intermittently remove ice from frozen food for quick microwave cooking. The defrost circuit may consist of a motor, switches, control board, and cam assembly in the oven.

dielectric   A material that serves as insulator. A dielectric material is used between the metal foil layers of the hv capacitor.

diode   Passes current in one direction and blocks current in the reverse direction. The hv diode has a cathode and anode terminal. The cathode terminal is always at chassis ground in the microwave oven circuits.

door assembly   The front door assembly opens so food can be inserted in the oven for cooking. The door assembly has hinges at one end and latch or hook switch assemblies to hold the door in place.

electron   The smallest electric charge. A negatively charged particle. Electrons flow from the cathode to the anode terminal at a high rate of speed in the magnetron.

fan blower   A fan motor to provide cooling of the magnetron and circulate the air in the oven areas.

ferrite   A compressed powdered iron material.

filament or heater   The heater or filament element heats up the cathode terminal to emit free electrons. The filament or heater terminals are at a high negative voltage in the oven circuit.

filament transformer   In some microwave ovens a separate transformer is used to provide voltage to the filament or heater terminals of the magnetron.

fuse   A chemical 15 amp fuse is found in all microwave ovens to protect the oven circuit.

fuse resistor   A service that senses an increase in current in the transformer secondary and opens the transformer primary circuit.

impedance   A combination of resistance and reactance, expressed in ohms.

inductance   The property of a circuit that causes a magnetic field to be induced.

interlock switch   The interlock switch prevents the operator from using the oven without having the door closed. Interlock switches may be called monitor and safety switches.

latch switch   An interlock switch located and controlled by the oven door.

leakage tester   A government approved survey instrument to measure radiation leakage of the front door, vents, and waveguide areas.

light switch   Controls the oven light when the door is closed or opened.

magnetron   The magnetron is a vacuum tube in which the flow of electrons from the heated cathode flow to the anode element. The speed of the electrons is controlled by a magnet and an electrical field. The magnetron produces very short electrical waves (microwaves). The frequency of these waves is 2,450 MHz.

oven light   Furnishes light to the oven cavity. Some ovens have two lights.

power cord   The power cord provides power from the ac outlet to the oven circuits.

power transformer   A large power transformer provides filament or heater voltage for the magnetron and high voltage to the hv circuits. A small power transformer provides low ac voltage to the control board.

radiation leakage   Rf microwave leakage that may occur around the door and vent areas.

relay   An oven relay provides power to the hv transformer and magnetron. The auxiliary relay energizes when the oven is placed in the cook cycle. The power oven relay may be controlled by a digital programmer circuit.

resistor   A device that limits the flow of current, providing a voltage drop. A high-megohm resistor may be found in the hv circuits. A 10 ohm resistor may be used to measure current in the hv diode circuit.

stirrer motor   A motor or blade that circulates the microwaves within the oven cavity. The stirrer blade may be driven by the fan or blower motor by a long drive belt.

temperature probe   When inserted into meat or other food in the oven cavity, the probe determines the temperature or cooking time.

thermal switch   When overheated, the thermal switch opens up the power-line voltage, preventing the magnetron from overheating.

timer   A device to control the amount of cooking time.

triac   An electronic switch to apply ac voltage to the hv transformer. Usually, the gate voltage of the triac is controlled from the control board.

voltage-doubler circuit   In a microwave oven, the voltage-doubler circuit consists of a silicon rectifier and hv capacitor. The high ac voltage from the secondary winding of the power transformer is applied to the voltage-doubler circuit where it is rectified and is always less than the doubled input voltage.

watt   The unit of electric power. A microwave oven may pull over 200 watts.

waveguide   A metal enclosure for the conduction or transmission of microwaves. The waveguide in the microwave oven is found between the magnetron and the oven cavity.

wavelength   The distance between corresponding points of two successive ac waves.

Index

A

alligator clip leads, 38-39
Amana microwave ovens, door replacement, 286
antenna, rotary, 27, 76
arcing, oven or transformer, 101, 178, 189-190, 201, 210-211, 324-325, 331

B

bake-broil control, 260-262
bleed resistor, 17
blower/blower motor, 17, 18, 218-221, 234-236, 320-321
blown fuses, 98, 100, 141-145, 293, 294, 303, 316-317
browning elements (see heating elements)
burning food, 101, 201, 322-325
buzzer or speaker, 17, 19, 258

C

capacitor, hv, 17, 19, 60, 61, 179-182, 350
Change/Cancel control buttons, 238
coils, 59, 61
components, 17-34
continuity tester, 40
control circuits and panels, 20, 63-64, 88, 237-272, 325-326
  bake-broil control, 260-262
basic operation steps, 239-240
buzzer control, 258
change or cancel, 238
continuity tests, 251
control board (DPC) circuitry, 243-252
cooking/heating cycle inoperative, 254
delay cooking, 238-239
display tubes, 258
humidity control and sensors, 252-254
installing new control board, 271-272
keyboard checks, 250-252
lightning damage, 269-270
memories, 237, 242-243
multivibrator circuit, 263-265
phase control circuits, 262-263
power level control, 237, 240-241, 254-257
power supply, digital programmer, 244
program check, 238
programmed test cycle, Whirlpool, 266
pulse oscillator circuit, 257
replacement and repair, 270-271
switching circuit, 263-265
temperature control, 238, 241-242, 258-260
temperature sensors, 254
time vs. temperature cooking, 237
timer control, 240-241
troubleshooting, general procedures, 265-268
variable power circuit, 262
warming features, 238
convection cooking circuits, 69-71
convection cooking inoperative or poor, 333
convection timer motor, 227-228
cook relay, 18, 85
cook switch, 18, 73
cooking test procedures, 50-53, 198-200
cooking/heating cycle, 71-77
  cuts out after startup, 148-152, 182, 202, 236, 311-313
  erratic or intermittent operation, 96, 98, 148-152, 201, 313-316
  inoperative, no cooking, 84-88, 97, 152-153, 158-159, 181, 201, 202, 254, 303-311
  test procedures, 50-53, 198-200
cover removal, 77-78
current testing procedures, 35, 44-49

D

dead condition, no oven functions, 82-84, 92-96, 100, 181, 201, 291-296
defrost circuits and problems, 20, 64-66, 90, 105
delayed cooking control, 238-239

digital multimeters (DMM), xii, 42-43

digital programmer circuit, 360-361

digital VOM, 41-42

diodes, high voltage circuits, 182-185

discharging hv capacitor, xii, 12, 17, 18, 46, 78

display tubes, controller circuits, 258

doors and door problems, 12, 21, 35, 36, 72-75, 100, 102-103, 284-289, 295, 327-330

E

electrical interference from oven, 9

electrical shocks and shock hazards, 10-16, 105, 321-322

electromagnetic waves, 1-3

electrostatic field generation, ac, 59

exploding items in oven, 210

exposure, microwaves, 35-36

extension cords, 12

F

fans and fan problems, 73, 99, 101, 214-219, 236, 298-301, 318-321

fluorescent tube (*see* also lights), 22

frequency, 1

fuse blown, 98, 100, 141-145, 293, 294, 303, 316-317

fuse puller, 39-40

fuses, xii, 13, 21, 29, 57, 58, 350

G

gas sensor, 23

grounding, 7, 12, 14, 106

H

Hardwick microwave ovens, door replacement, 286

heating elements and problems, 21, 72, 105, 161-165

heating/cooking cycle (*see* cooking/heating)

high-voltage circuits, 61-62, 167-190

capacitors, 179-182

diodes, 182-185

magnetron, 167-169

test and service procedures, 185-189

transformer, 173-179

voltage-doubler circuit, 169-172

hinges, door, 102

Holaday Model HI-1500 leakage testing survey meter, 275-276

hot spots, 101, 189-190, 201

hum or noisy operation, 98, 101, 318-319

humidity control and sensors, 23, 252-254

I

inoperative oven (*see* dead condition)

installation, 7-10

interference from oven, electrical, 9

interlocks, xii, 12, 21, 35, 36, 72, 85, 86, 88, 99, 109-120, 350

adjustments, 118

replacement, 115-120

safety switch systems, 118

troubleshooting guide, 117-118

intermittent operation (*see* cooking=heating cycle, intermittent)

isolation of problems, 82

L

leakage testing, xiii, 13, 35, 36, 53-56, 107, 273-289, 321-322

customer leakage tester, 282-283

location of possible leaks, 277-280

measuring precautions, 277

Norelco test procedure, 276-277

procedures, 280-282

radiation leakage monitors, 275-276

record keeping, 283-284

safety precautions, 273-275

test equipment, 275

waveguide assembly replacement, 289

light switch, 23

lightning damage, 269-270, 330

lightning protection, 13, 16, 13

lights, 25, 73, 75, 88, 90, 103, 136-138

intermittent or out, 146-148, 296-297, 300-301, 332

remain on, cannot be turned off, 316

literature, service, 345

low-voltage circuits, 57-61, 139-166

fuses, 141-145

heating elements, 161-165

intermittent cooking, 148-152

lamp triac testing, 158

lights on, no cooking, 158-159

lights, intermittent, 146-148

no cook cycle, 152-153

reversed triac connections, 157-158

shutoff inoperative, 153-157

temperature probes, 158-161

thermal cutout switch, 148-152

troubleshooting procedures, 139-141

varistor test, 146

M

Magnameter, xiii, 44-49

magnetic field storage, coils, 59

magnetron tube, 3-7, 13, 23-25, 36, 167-169, 191-212, 350

arcing in oven, 210-211

cooking test, 198-200

cutaway view, 192

exploding items in oven, damage to, 210

failure of unit, 200-204

internal parts diagram, 192

magnetic tools warning, 207

ohmmeter testing, 196-198

ordering new tubes, 210-212

overheating ovens, 209-210

removal procedures, 204-207

repair vs. replacement costs, 209

replacement procedures, 207-209

thermal protector switches, 202-204

VOM testing, 194-195

warranty replacement, 212

meat probes (*see* temperature probes)

memory circuits, 237, 242-243

microwave leaks (*see* leakage testing)

monitor or safety switches, 25, 36, 72, 74, 120-124

motors, 213-236

blower, 218-221

blower, auxiliary, 234-236

convection timer, 227-228

fan, 214-219

stirrer, 221-225

timer, 225-227
turntable, 214, 228-232
vari-motor, 232-234
multivibrator circuit, 263-265

N

Narda 8100 radiation monitor, 55
noisy operation or hum, 98, 101,
  236, 318-319
Norelco microwave ovens
  body layout, 355
  diode testing, 184-185
  door replacement, 286-287
  fan motor replacement, 219-221
  interlocks, 114
  leakage testing procedures, 276-
    277
  magnetron tubes testing, 197
  magnetron tubes thermal pro-
    tector, 202-204
  monitor switches, 124
  on/off switches, 127
  parts lists, 356
  schematic diagram, 81
  stirrer motor, 221-224
  wiring schematic, 354

O

ON/OFF switch, 125-127
operation of microwave ovens, 7-
  9, 239-240
outlet receptacle, polarity test, 13,
  15, 13, 106
oven relay, 127-130
overheating, 101, 209-210, 322-
  325

P

Panasonic microwave ovens, 28
  bake-broil control, 260-262
  buzzer control circuit, 258
  capacitor testing, 182
  control board (DPC) circuits,
    244-246
  convection circuits, 70-71
  cook testing, 50, 199
  defrost circuits, 65-66
  digital programmer circuit, 360-
    361
  digital programmer power sup-
    ply, 244
  display tubes, controller cir-
    cuits, 258
  exploded view, 359
  heating elements, 165
  high voltage circuits, 170-172

humidity control and sensors,
  252-254
interlocks, 118
keyboard checks, controller
  circuits, 250-252
leakage testing, 56
magnetron tubes replacement,
  209
magnetron tubes testing, 197
oven cavity construction, 188-
  189
positive lock connector, 177-
  178
power level controls, 254-257
power relay control, 131-134
pulse oscillator circuit, 257
schematic diagram, 357
stirrer motor, 224-225
temperature control, 66-69,
  258-260
temperature probes, 159-161,
  254
transformer, 174-175
variable power circuits, 262
wiring diagram, 358
parts list, 356
parts suppliers, 341-351
phase control circuits, 262-263
polarity test, outlets, 13, 15, 106
popping noises, 98, 182
positive lock connector, 177-178
power level control, 237, 240-241,
  254-257
power line surges, 13, 16, 13
power outages, no resumption of
  operation, 331
probe jack, 25
program check control, 238
pulse oscillator circuit, 257

Q

Quasar microwave ovens, door
  replacement, 288

R

radiation leaks (*see* leakage testing)
rectifiers, 25-26, 350
reflection, microwave, 3
relays, 18, 26-27, 75, 85, 127-130
  contact layout, 130-134
resistor, bleed, 17, 27
resonant circuits, 5-7
returning parts to manufacturer,
  347-348
rotary antenna, 27-28, 76

S

safety during operation, 7-9, 335-
  339
safety during service, 10-16, 273-
  275, 335-339
Samsung microwave ovens, door
  replacement, 288
schematic diagram, 81, 357
screwdrivers, 37-38
selector switch, 138
sensor probe, 69
sensors, 23, 252-254
service forms, 350-351
service literature, 345
Sharp microwave ovens, door
  replacement, 289
shocks and shock hazards, 10-16,
  105, 321-322
shutoff inoperative, cannot turn
  oven off, 153-157, 319-320,
  331
Simpson 380 series leakage tester,
  55-56
slow cooking speed, 97, 201, 236,
  308, 313-316, 330, 332
smoking from oven, 322-325
sparks, oven or transformer, 101,
  178, 189-190, 201, 210-211,
  324-325, 331
speaker (*see* buzzer)
start switch, 125-127
stirrer motor/shield, 27-28, 221-
  225
stop switch, 28, 125-127
substituting parts, 348-350
suppliers, 341-351
switches, 18, 109-138
switching circuit, 263-265

T

television, electrical interference
  from oven, 9
temperature control circuits, 66-
  69, 238, 241-242, 258-260
temperature fuse, 29-30
temperature probes, 158-161, 333
temperature sensors, 254
test equipment and tools, 35-56
test equipment attach points, 36-
  37
thermal cutout switch, 29, 148-
  152
thermistors, 29-30
thermostat, 30-31

timers and timer problems, 30-32, 104, 134-136, 225-227, 240-241, 297-298, 318-319
tools and test equipment, 35-56
transformers/transformer problems, 32, 89-90, 173-179
transistors, 43
triac, testing and troubleshooting, 32-33, 75, 88, 92-93, 105, 107-108, 153-158
   lamp, testing, 158
   reversed connections, 157-158
troubleshooting, general procedures, 79-108
   charts for, 90, 93-96

turntables and turntable problems, 33, 105, 214, 228-232, 236, 301-303

V

vari-motor, 33, 232-234
vari-switch, 33
variable power circuits, 262
varistors, 33, 146
volt-ohmmeter, 41
voltage testing, 12, 14, 35, 44-49
voltage-doubler circuit, 169-172

VTVM, 43-44

W

warming features, 238
warranty parts, 345-347
water test (cooking test) procedures, 50-53
waveguides, 34, 36, 289
wavelength, microwave, 3
Whirlpool microwave ovens
   controller circuits, programmed test cycle, 266
   cook testing, 50-53, 199-200
wiring, 7-9, 80
wiring diagram, 28, 358